COMETS
Speculation and Discovery

Nigel Calder

DOVER PUBLICATIONS, INC.
New York

Copyright

Published in Canada by General Publishing Company, Ltd., 30 Lesmill Road, Don Mills, Toronto, Ontario.
Published in the United Kingdom by Constable and Company, Ltd., 3 The Lanchesters, 162–164 Fulham Palace Road, London W6 9ER.

Bibliographical Note

This Dover edition, first published in 1994, is a slightly revised and corrected republication of the edition, published by Penguin Books (Harmondsworth, Middlesex, England) in 1982, of the work originally published by BBC Books, London, 1980, under the title *The Comet Is Coming!: The Feverish Legacy of Mr Halley*. The author has written a new Preface specially for this edition.

Library of Congress Cataloging-in-Publication Data

Calder, Nigel .
[Comet is coming]
Comets : speculation and discovery / Nigel Calder.
p. cm.
Originally published: The comet is coming. Harmondsworth, Middlesex, England ; New York, N.Y., U.S.A. : Penguin Books, 1992, c1980.
Includes bibliographical references and index.
ISBN 0-486-27879-4 (pbk.)
1. Comets. 2. Halley's comet. 3. Halley, Edmund, 1656–1742. I. Title.
QB721.N54 1994
523.6—dc20 93-38506
CIP

Manufactured in the United States of America
Dover Publications, Inc., 31 East 2nd Street, Mineola, N.Y. 11501

PREFACE TO THE DOVER EDITION

Halley's Comet came in 1986 and went away again, leaving the world better informed about comets, but still prone to extraordinary speculations about them. While the original title of this book, *The Comet Is Coming!*, is now an anachronism, the subtitle, *The Feverish Legacy of Mr Halley*, is not. The book remains valid as an account of the endless tussles between solemn science and the crazy ideas that sometimes turn out to be correct. This preface does no more than point out a few new twists to the story.

Five spacecraft intercepted *Halley* in March 1986: the Soviet Vega 1 and Vega 2, the Japanese Sakigake and Suisei, and the European Space Agency's Giotto, which went deepest into the comet's head. Information from the Vegas and from NASA's tracking stations enabled Giotto to approach within 600 kilometres of the nucleus on 14 March 1986. It passed on the sunward side, and not the tailward side indicated in the diagram on page 143. Up to the moment when *Halley*'s dust damaged it badly, the European spacecraft took glorious close-up pictures of the comet's nucleus and its luminous fountains of dust and gas.

The nucleus was extremely black, with no signs of ice visible at its surface. It looked more like a dirtball than a snowball. Chemical instruments in the spacecraft nevertheless detected plenty of water vapour escaping from the comet, as expected by the fashionable "dirty snowball" theory (Chapter 5). In the opinion of Uwe Keller of Katlen burg-Lindau, the chief analyst of the Giotto pictures, a comet should not be thought of as a ball of ice impregnated with dust grains. He thinks the non-icy materials give the comet its structural integrity.

The water of *Halley* seems indistinguishable from the Earth's in its atomic composition. This discovery by the visiting spacecraft has comforted those scientists who like to say that the young Earth acquired its oceans and atmosphere from impacting comets. The Vegas and Giotto also detected, emerging from *Halley*, many small dust grains rich in elaborate compounds of carbon. They have prompted a new round of speculations about the link between comets and life.

Most experts still scorn the theory that the comets themselves contain living organisms (Chapter 6). They treat more seriously the possibility that carbon compounds raining on the Earth from comet tails gave essential flavouring to the primeval soup in which life supposedly began. Jochen Kissel of Heidelberg, whose dust-analysing instruments were carried in Giotto and the Vegas, goes further. He hypothesizes that grains delivered by a comet impact reacted chemically with the soup of the young Earth in just the ways required to initiate life. Earth remains our mother, in Kissel's scheme, but an anonymous comet is our father.

Over the death of the dinosaurs (Chapter 7) controversies still rage. Just when the physicists were beginning to persuade the sceptical fossil-hunters that the impact of a comet or asteroid was indeed responsible for clearing the Earth of large reptiles sixty-five million years ago, other scientists interrupted with another theory. They put the blame on enormous volcanic outpourings in India, which happened at around the same time. On the other hand, the extraterrestrial explanation is bolstered by the discovery of a giant impact crater of just the right age, buried under the coast of Mexico.

The proposal in Chapter 8 that human beings should defend the Earth against threatening comets and asteroids was seriously meant, though expressed lightheartedly. The politicians are catching up with the idea. NASA studies, ordered by the U.S. Congress, have led to a set of proposals called Spaceguard. It envisages a more intensive search for menacing objects, and the development of means to deflect or destroy them.

Halley's traditional ability to surprise the Earthlings has not been neutralized by the power of modern science. Five years after its visit to the Sun, *Halley*'s nucleus was an extremely faint, bare object. At a distance of more than 2 billion kilometres, in the cold realm far beyond the orbit of the planet Saturn, it was only just visible to powerful telescopes. By conventional cometary wisdom, it should have settled down to tailless quiescence until its next visit to the Sun.

Then, in February 1991, *Halley* suddenly erupted. It generated a brand-new dust cloud 200,000 kilometres wide, and brightened 300-fold. The event set off a frenzy of speculation among comet experts. Some said a large meteorite had hit it. Others theorized about an internal explosion, triggered perhaps by a change in the crystal structure of the comet's ice. And older forms of comet fever reappeared in advertisements in London newspapers, proclaiming that *Halley* had changed its orbit and was rushing back towards the Earth—"Unexpectedly! Now!"

For more detailed updating, the reader is referred to my book *Giotto to the Comets* (London: Presswork, 1991). It also tells how the European spacecraft survived to visit a second comet, *Grigg-Skjellerup*, in July 1991.

NIGEL CALDER

CONTENTS

1618

1910

1577

Tokens of comet fever. In former times medals were often struck to give thanks for deliverance from the evils of the comet; more recently they have commemorated the return of Halley's Comet.

CONTENTS

AUTHOR'S NOTE

As usual it is my pleasure to thank the BBC for its support in this project and the scientists who gave so freely of their time and advice. Without slighting many others who have contributed indispensable information, I should like to acknowledge special help and insight from the following: Walter Alvarez, Jerome Bruner, David Dale, Freeman Dyson, Owen Gingerich, Michael Hoskin, David Hughes, Tao Kiang, Raymond Lyttleton, Brian Marsden, Marcia Neugebauer, Ray Newburn, Ernst Öpik, Giampaolo Pialli, Simon Schaffer, Zdenek Sekanina, Jan Smit, George Wetherill, Fred Whipple and Donald Yeomans. They are not, of course, responsible for any errors nor for my less than reverent approach to their subjects. I should also like to thank Enid Lake and Peter Gill at the Royal Astronomical Society's Library, for their help in finding books and journals, old and new.

1618

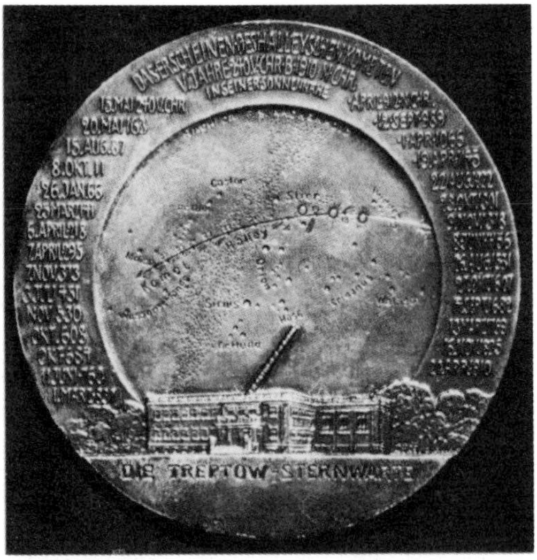

1910

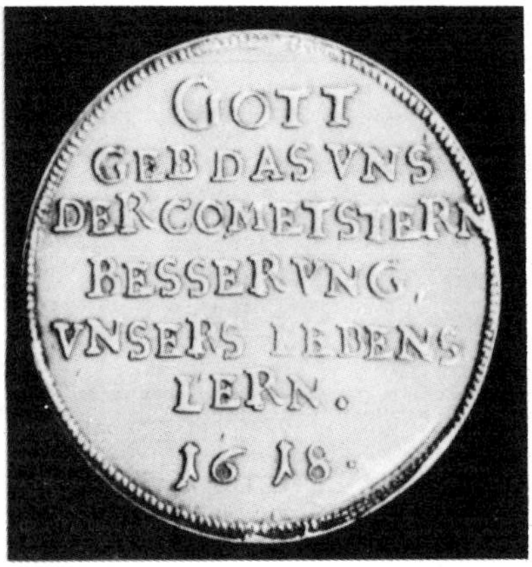

1577

Tokens of comet fever. In former times medals were often struck to give thanks for deliverance from the evils of the comet; more recently they have commemorated the return of Halley's Comet.

1.
TELEGRAMS FROM THE GODS

Soothsayers and fiction-writers have a case: one day the Earth will collide with a bright comet or its dark corpse and the result will be world-wide mayhem. It has happened before. But the odds against a repetition in anyone's lifetime are very long indeed, so worry sooner about the effects of comets on mental health. Smudges of light move against the background of stars in what the comet lovers in our midst might say was a stately fashion (I would call it slovenly) and they infect the imagination of laymen and astronomers alike.

A comet travels faster than a spaceship, at many kilometres per second, but because it is far off it spends weeks inching its way across the sky. With a bright, roughly round head and a hair-like tail, a typical comet looks like a hurtling missile with streak-marks in its wake; you have to make a mental effort to realise that the object may be moving in any direction and the tail is never trailing straight behind it. Comets are sometimes confused with their impetuous kin, the meteors and meteorites that dash into the Earth's upper air from outer space and either burn up as 'shooting stars' or reach the ground as incandescent lumps of iron, stone and tar. The astronomers' term for the advent of a comet is an 'apparition', and its connotation with wraiths is wholly appropriate.

The most provocative apparition of all is now upon us, because the comet we call *Halley* is slanting in towards the Sun. It began its relentless countdown in 1948, when it faltered at the top of its trajectory, at thirty-five times the Earth–Sun distance. Since then it has been falling back, gathering speed, and in the 1970s it drew nearer than Neptune, the most remote of the great planets of the Sun. In 1977 the first of the large telescopes looked out for it, in the constellation of Canis Minor, but there was little hope of spotting the comet, small, faint and tail-less, as early as that. After further unsuccessful attempts, astronomers resigned themselves to not recovering *Halley* until the early 1980s. During 1981–4 the comet's path passes in front of the Milky Way, making it hard to see. In 1985, it will swoop into the heart of the Solar System and deploy its wanton tail. It will swing around the Sun early in 1986, being at its most visible in the first few months of that year. Then it will climb away and few of us will see it coming back, yet again, in 2061.

Every seventy-six years or so, *Halley* punctuates history like an exclamation mark in the sky. On its last return in 1910 the more spectacular 'Great January Comet' upstaged it but, as the brightest comet that reappears frequently, *Halley* exerts a special grip on the human mind. Although the present apparition will be a meagre one, to say so will only boost the sales of binoculars and amateurs' telescopes. Professional astronomers know perfectly well that comets are a swindle: small, lightweight objects that look awesome because they generate voluminous heads and tails. They are striking astronomical sights and deserve an explanation, yet even when their tenuous nature has been fully disclosed they continue to cause astonishment and excitement out of all proportion to their substance. The coming of *Halley* means that we are due for a bout of comet fever.

Comets drive people dotty. Like the emperors and priests who used to tremble when they appeared, some members of the public are still eager to be duped by charlatans selling protection against the evil influence of comets, or pamphlets proclaiming the end of the world. But eminent scientists, too, are vulnerable to the fever and concoct their most bizarre theories around comets. At one time Noah's Flood was blamed on the impact of a comet, and in the 1980s there is a solemn assertion that *Halley* causes influenza. While I shall not skimp Sir Fred Hoyle's speculations about such fevers afflicting the body, the psychiatry of comets seems less debatable.

A few prophylactic statements about where they and we operate in the cosmic scheme may be appropriate at the outset. The universe is large and violent, but we survive because our wagon (the Earth) is hitched to a bright but humdrum star (the Sun), far from the tumultuous centre of our Galaxy (the Milky Way). Other stars in our neighbourhood scurry about, but only two interstellar matters are relevant to our parochial story: passing stars occasionally encounter comets wandering far from the Sun, and the Sun sometimes conducts all its family through dark, diffuse clouds of interstellar dust.

That family, the Solar System, consists of two significant planets, Jupiter and Saturn, which resemble small, cool stars. In addition there are two fairly large icy planets, orbiting around the Sun far beyond Saturn: Uranus and Neptune. Reading inwards from Jupiter, there are four other planets, stony and much smaller: Mars, Earth, Venus and sweltering Mercury, closest to the Sun. After those, you can fill in further small planets like Pluto, Chiron, and the asteroids that circle between Jupiter and Mars; assorted moons that in some cases rival the stony planets in size; the rings that encircle Saturn, Jupiter and Uranus. Comets come close to the bottom of the list, but they obtrude because they jaywalk across the paths of the planets.

Few professional astronomers devote themselves full-time to the study of comets. The capital of their small kingdom is at an astrophysical observatory in Cambridge, Massachusetts, where a handful of experts record the sightings and orbits of all comets and lead the scientific efforts to understand them. In the theory now prevalent in Cometsville, a comet is a cosmic sorbet: a dirty snowball that comes tumbling out of the freezer of twilight space, far from the Sun. In commending this description as an anchor in the storm of other strange hypotheses I do not mean that you can be quite sure of it, at least not until a spaceborne robot or an astronaut lands on the supposed snowball in the heart of a comet.

In the warmth of the inner Solar System a comet releases clouds of vapour and dust that form the glowing head and then leak into the tail, which is the cosmic equivalent of an oil slick. Pieces of the dust later hit the Earth, as meteors. A few survivors among the comets evolve into menacing lumps of dirt in tight orbits around the Sun. For these reasons comets are, in my opinion, best regarded as a conspicuous form of sky pollution.

A description of comets by Maurice Dubin of the US National Aeronautics and Space Administration is candid:

> As the source of meteor streams and meteors in general they are presently viewed as a third-rate cosmic population lacking any influence on the goings-on of this world. . . . Perhaps their study may lead to unexpected discoveries despite the insignificance of these bodies.

To hold their own at scientific gatherings, Dubin and other students of the 'third-rate' objects like to say that comets will shed light on the primordial chemistry of the Solar System which gave the Earth and the other planets their present character; they may also help in understanding the origin of life. Those are quite reasonable hopes. But as a contamination of the Solar System, from which meteors are the least objectionable end product, comets continue to influence 'this world' less benevolently than that.

They violate the elegant order of the heavens, and the response of nations without telescopes has often been astrological and pessimistic. This is the primitive form of comet fever. When you have established their true paths through space, a secondary affliction develops: an urge to see as many comets as possible. Finally comes the knowledge that cometary dust enters our atmosphere, and that comets themselves must sometimes strike the Earth's surface. These direct contacts with comets give rise to feverish theory-spinning, which has to be coolly assessed in order to diagnose any real dangers and formulate countermeasures.

Our generation may be better informed than previous ones but it is not immune to error. Indeed there is, at first sight, little to choose between some present-day propositions about comets and those that conceited folk mock as naïveties of their forefathers. Because *Halley* and its like are ostentatious and elusive, they create a no-man's-land of speculation where crazy ideas may be fantasies or brilliant insights. The interplay of imagination and evidence in this annexe of astronomy says a good deal about the nature of science, and describing it may help to inoculate the more reasonable sections of the public against the predictable nonsense of *Halley*'s return.

These apparitions are dangerous because human beings make them so. Comets kill people by self-fulfilling superstition, when those who read them as telegrams from the gods or the Devil turn in panic to homicide or suicide. Compared with all the elaborate contrivances about planets in the zodiac, the astrology of a comet is relatively simple: it is a disorder of the heavens 'importing change of times and states', as William Shakespeare put it, and is bad news, especially for eminent persons. In Roman times, no one was a more eligible victim than the Emperor, and when a bright comet appeared in AD 60, or thereabouts, the public knew what it meant. The historian Tacitus wrote: 'As if Nero were already dethroned, men began to ask who might be his successor.' Safety for the Emperor lay in using his most prominent subjects as omen-conductors and letting the message in the sky be fulfilled in their deaths rather than his.

The astrologer Balbillus assured the young Nero that it was

customary for monarchs to deflect the wrath of heaven in that way – potent advice, you may think, for the man who murdered his mother, two wives and most of his family, and set Rome on fire. Nero took no chances as another historian, Suetonius, related: 'Nero resolved on a wholesale massacre of the nobility. . . . All children of the condemned men were banished from Rome, and then starved to death or poisoned.' The policy worked like a charm. Nero survived that comet by several years; even a visit by *Halley* in AD 66 failed to get rid of him, and eventually he committed suicide at the age of thirty-two. Meanwhile public expectations had been gratified and eminent folk had indeed perished on account of the comet.

The Incas of Peru regarded comets as intimations of wrath from their Sun-god Inti, and they presumably sacrificed a few more children to calm him when *Halley* passed by in 1531 shortly before Francisco Pisarro conquered them. In twentieth-century Oklahoma, at the apparition of *Halley* in 1910, the sheriffs arrived just in time to prevent the sacrifice of a virgin by demented Americans calling themselves Select Followers. By then the evil was more liberal. The march of democracy undermined the belief that comets were targeted on the high and mighty. Many people without any claims to kingship killed themselves to escape the luminous horrors of the night, thus ensuring that the comet had its due. And if astrologers might judge a motley collection of Italian and Spanish suicides to be unworthy of the portent, they could congratulate themselves that Edward VII, King of the Empire on which Sun and comet never set, perished promptly at his cosmic cue – from the comet complicated by bronchitis.

Did not the sungrazer *Pereyra* signal President Kennedy's murder in Dallas in 1963, and what was the less than fatal *Kohoutek* of 1973–4 but a symbol of Nixon's disgrace, fizzling out like a deleted expletive? Cometomancy is easy, especially if you bend the rules. When the bright comet of 44 BC appeared several months late, *after* the murder of Julius Caesar, the portent theory was temporarily set aside and the comet was said to be the victim's soul ascending into heaven. By an oversight no comet at all appeared in AD 814, to herald the death of Charlemagne, Emperor of Europe. Yet to the chroniclers this precursor of the death of kings was as natural as a pregnancy before birth; so, undismayed by want of facts, they duly recorded that a comet came. Modern astronomy can exonerate the scholars of any grave lie: there is always a comet or two in the offing, whether or not we see them.

The Purple Mountain Observatory at Nanking in China, with instruments ancient and modern. A careful watch for comets continued here for many centuries, and recently the observatory has made its mark again with the discovery of new comets.

If only Edmond Halley had worked for the Emperor of China his discovery of the repetitive behaviour of a comet might have been hushed up. With hindsight we know that his comet was one among four that appeared in AD 837 – a prodigality that caused world-wide dismay. The Chinese court classified astronomy as secret because traitors in the imperial observatory, or freelance astrologers, might give aid to rebels seeking to overthrow the Emperor. Astronomers in China filled the same kind of role, wanted yet mistrusted, as makers of nuclear weapons do today. Their work was always liable to censorship but the edict of AD 840, as quoted by Joseph Needham, the historian of Chinese science, was particularly sharp: 'The astronomical officials are on no account to mix with civil servants and common people in

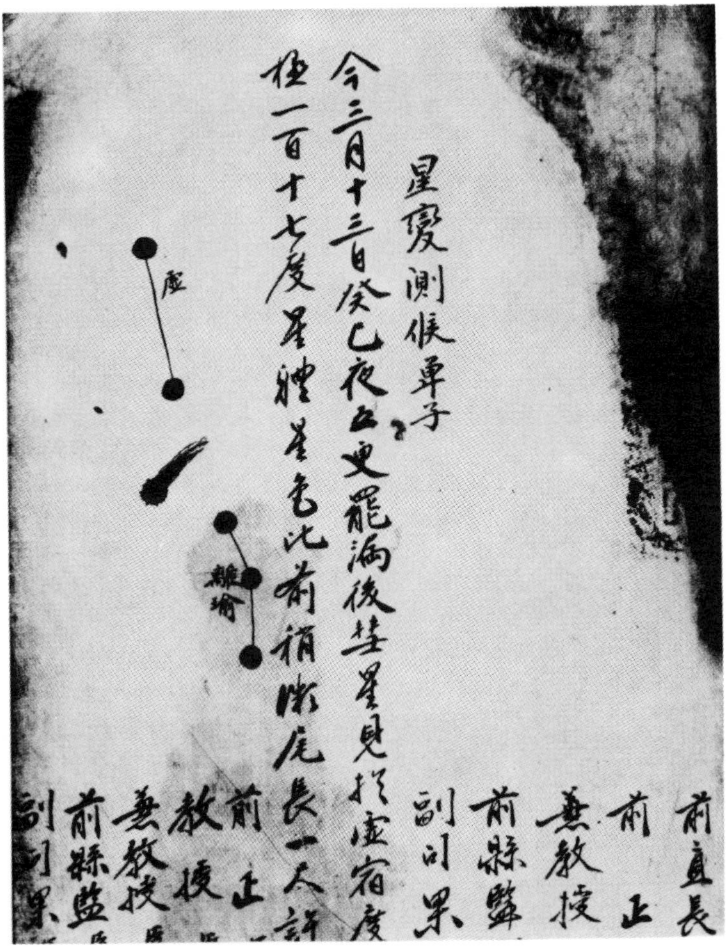

A Chinese record of a comet – in fact Halley's Comet at its first predicted return, in AD 1759. In this comparatively recent work the astronomer was continuing a tradition going back more than 3000 years. The 'report sheet' says that on 13 March, after rain, the comet is 117 degrees from the pole, and the tail has shortened a little.

general.' Later it became a capital offence to study mathematics in China without royal permission.

A legend tells of two astronomers, Hi and Ho, who were too drunk even to notice an eclipse of the Sun, and the Emperor had their heads cut off. This early attempt to check insobriety among astronomers is an indication of the importance attached to their craft in ancient China. One of the principal observatories still stands on the Purple Mountain beside Nanking, although the precise site of the ancient observatory is now occupied by a weather research institute. When the Jesuit scholar Matteo Ricci was touring China in 1600, Nanking was the capital and Ricci was impressed by what he found at the Purple Mountain. Needham translates Ricci's approving description.

On top there is an ample terrace, capitally adapted for astronomical observation, and surrounded by magnificent buildings erected of old. Here some of the astronomers take their stand every night to observe whatever may appear in the heavens, whether meteoritic fires or comets, and to report them in detail to the Emperor.

The instruments were 'carefully worked and gallantly ornamented', with circles divided into $365\frac{1}{4}$ degrees. The Chinese knew four thousand years ago that the Sun takes just that many days to cycle around the sky in the course of the year; the western circle of 360 degrees comes from a cruder Babylonian approximation to the number of days in the year.

The Emperor's power derived from keeping the calendar, so his chief astronomer and his staff were much concerned with the Sun, the Moon and the changing seasons, and with the planet Jupiter, the motions of which define a cycle of twelve years. They also had to issue long-range weather forecasts and make certain that eclipses did not kill off the Sun for good. And they needed to know every permanent star and nebula visible to the naked eye, so that they could spot any comets and other strangers instantly. For thousands of years the Chinese astronomers kept watch on their chilly terraces, so of course they saw hundreds of comets wherewith to frighten themselves and their Emperors.

Facing: Early portraits of Halley's Comet. The apparition of 1066 figures in the Bayeux Tapestry (above) as one of the key events leading to the Norman Conquest of England; the public shows the consternation caused by a conspicuous comet. And Halley's Comet 1301 appears in the Adoration of the Magi *(lower picture) in the Arena Chapel at Padua, which Giotto painted not long after seeing the comet; it served as the Star of Bethlehem. As a result, the European Space Agency's mission to Halley's Comet in 1985–6 is named Giotto.*

They left valuable records for their modern successors, the most famous being the Chinese observation of a 'new' star in AD 1054, now known to have been the exploding star whose remnants are visible today as the Crab Nebula. They may also have seen, unwittingly, the creation of a black hole, one of those bottomless pits in the sky that seem to be an inescapable consequence of Albert Einstein's theory of gravity. In October 1408 another 'guest star' burned brightly for forty-four days in much the same direction as the X-ray star Cygnus X-1 now lies – an object strongly suspected of containing a black hole that was made during the explosion of a massive star. But the Chinese themselves were stronger in astrology than in astrophysics.

The motives behind all their vigilance are best revealed by the robot back-up system that they developed so that their Emperor's sex life need not be halted by bad weather. The Chinese invented clockwork to drive armillaries that represented the motions of Sun, Moon and planets in relation to the stars. It was essential to know the precise configuration of the sky at the time of conception of a prince, so that the prognoses could be made, and with their cunning machinery the astrologers could allow events in the

ISTI MIRANT STELLA
HAROLD

ASTRONOMICVM CAESAREVM

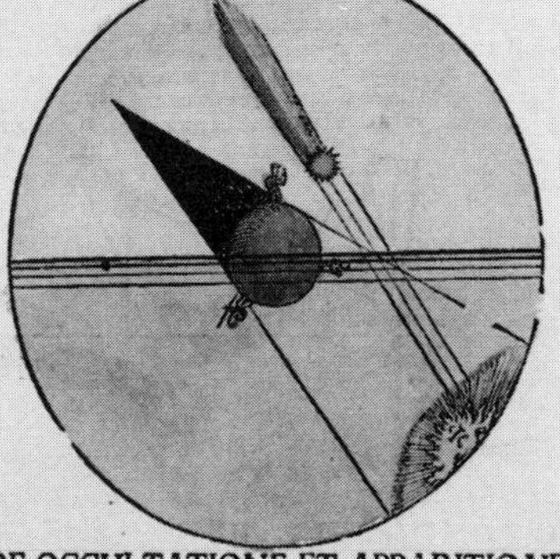

DE OCCVLTATIONE ET APPARITIONE Cometæ. Caput decimumsextum.

TAMETSI Cometes 13 augusti die mihi cõspectus sit primo, non desunt tamen qui eundem se vidisse sexto septimoq; affirment, verè autem omnibus ab occidẽte paucis ab oriente visus est, & profecto fieri non potest, quin ita factum sit, licet non omnibus apparuerit, non enim ita multis post ortum suum diebus videri mane desit, ea de causa, quod subinde magis ac magis cosmicum in ortum festinarit. Nam die 18, vt videre in sequentibus est, oriri cum ipso Sole cepit, quocirca propinquor erat iam Soli 13 die quam vt cerni potuerit, ortum parante iam Sole. Inde factum est, vt plerisq; imperitioribus alius ab eodẽ cometa fuisse putaretur, quasi duo forent, vnus in oriente, alter in occidente, nescientibus tam in ortu quàm in occasu apparere posse, non secus atq; stellã aliam. Id quod etiam Mattheo Palmerio Florentino imposuit in cronicis suis asserẽti binos in Ianuario cometas fulsisse, anno post Christũ 729, aliumq; Solem anteisse, alium vero subsequutum, cum reuera non duo, sed vnus & idem fuerit, nunc ante nunc post Solem, ea qua dixi ratione, lucens. Sed ad rem, Sol cum 29 Leonis gradum possederet, occidebat hora post meridiem 6 mi. 54, Cometa vero hora 9 mi. 55, si declinatio eius ascensioq; recta considerentur. Vnde liquet Solem fuisse sub horizonte horis 10 mi. 12, & Cometam horis 4 mi. 10, cuius idcirco ortus hora 2 mi. 10 post noctis medium esse debuit. Cum vero de ortu & occasu cometæ loquor, non subaudias volo, mihi visum per instrumenta eo tempore quo horizontem contigerit Cometa, illud enim abundè declinatio ascensioq; recta eius semel cognita suggerunt, & si vero per instrumentum id essem conatus, confieri tamen nequaq̃ potuisset, quoties enim horizonti propinquior fieret 3 vel 4 gra. interuallo, flamma eius extingui omnino videbatur, adeo, vt aliquandiu à me sub nubila condi creditus sit, donec admoneret Plinius libro naturalis historiæ 2 cap: 25, Cometas in occasura cœli parte nunquàm esse, id est, nunquàm apparere. Is enim genuinus Plinii intellectus est, ni fallor, quem eò libentius ego quoq; admitto, quoniam ita oculis esse compererim. De vsu meteoroscopii nihil amplius agã, sed post finem cometæ, motum quoq; diurnum per meteoroscopium inuenire docebo, sufficiat autem hanc vnam obseruationis methodũ ita fusè, & quasi per membra tradidisse, in sequentibus enim nil preter obseruationes meras per meteoroscopium consequutas proponem'. Figuræ præterea sequentes ostendunt altitudines Cometarum dumtaxat, supra horizontem existentium tempore quo Sol occumbit, similiter quã profundus Sol extiterit sub finitore cometa eundẽ scandẽte.

Sententia Plinii de occasu Cometæ.

SOLIS ET COMETAE ORTVS OCCAsusq; qui contingunt 13 die Augusti.

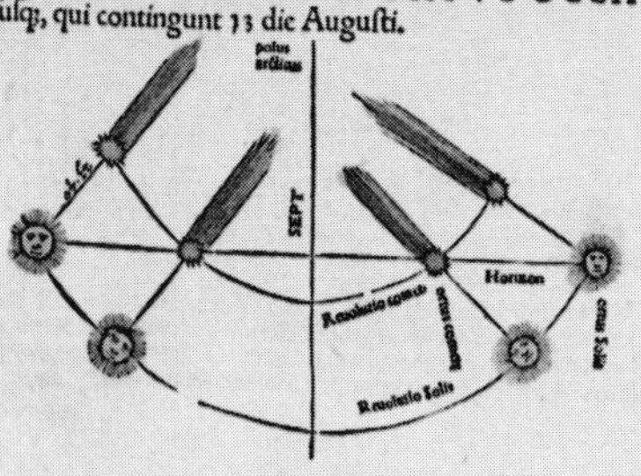

DECIMO QVARTO AVGVSTI DIE secunda Cometæ obseruatio peracta est.

¶ Altitudo Cometæ supra horizontem gra. 8 mi. 29
¶ Azimuth Cometæ ab occasu Septen. versus gr. 45 mi 22
¶ Altitudo extremitatis caudæ graduũ 23 mi. 18
¶ Azimuth extremitatis caudinæ Septentrio. gra. 57 mi 38

Hæc sequentia ex obseruatione consurgunt.

Latitudo Cometæ gra. 23 m̃ 2. Locus Cometæ gra. 23 mi. 39 ♌.
Declinatio Come gr. 35 m̃. 12 Sept. Ascẽsio recta Co. gr. 155 m̃ 5
Co. Mediat cœlũ ho. 11 añ. Amplitudo. or. & occ. 60 gr. 27 m̃ Sep.
Occa. Come. hora 9 m̃ 32 post me. Come. occidit cũ 23 gra. ♍
Distantia Come. à Sole gr. 23 m̃ 40. Ortus Come. hora 2 mi. 28.
Loc' extremi. caudæ gr. 39 m 38 ♌. Latitu. extre. gr. 39 m̃ 45 Sept.

Situs Cometæ occasus tempore. Situs Cometæ ortus tempore.

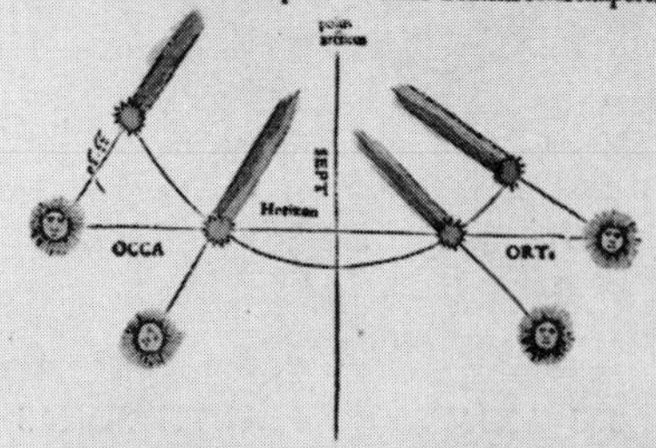

DECIMO QVINTO AVGVSTI OBseruatio tertia facta est.

¶ Altitudo Cometæ supra horizontem gra. 9
¶ Azimuth Cometæ ab occ. versus Sept. gra. 41 mi. 22.
¶ Altitudo extremitatis caudæ supra horizon. gra. 39 mi. 8
¶ Azimuth huius extremitatis gra. 50.

Hæc autem obseruatio collegit ea quæ sequuntur.

Latitudo Come. gra. 22. Locus verus Come. gra. 24 mi. 29 ♌.
Decli. Come. gr. 33 m̃ 19 Sep. Ascen. recta Co. 159 gra. 20 mi.
Latitudo extremitatis caudæ ab ecliptica gra. 34 mi. 22
Locus verus in ecliptica extremæ caudæ 23 gra. 2 mi. ♌
Distan. à Sole gra. 23 mi. 31. Arcus diur. Come. horæ 18 mi. 34
Occa. Come. ho. 9 mi. 44 post. Or. Come. hora 3 mi. 10 ante.
Hoc die Com. heliace accidit, vt nõ amplius ante Solis ortũ cerneret.

Situs Cometæ occasus tempore. Situs Cometæ ortus tempore.

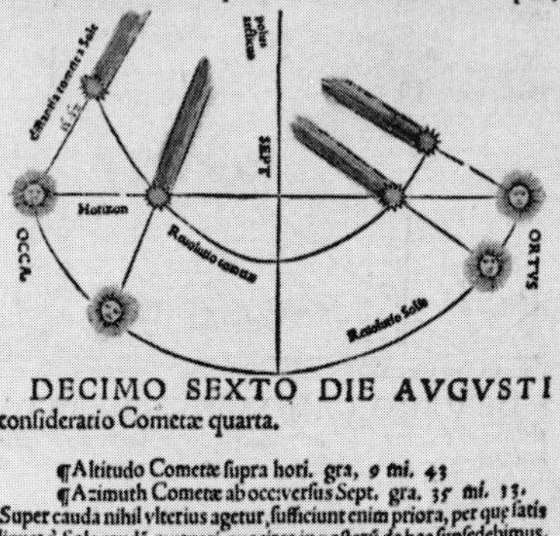

DECIMO SEXTO DIE AVGVSTI consideratio Cometæ quarta.

¶ Altitudo Cometæ supra hori. gra. 9 mi. 43
¶ Azimuth Cometæ ab occ. versus Sept. gra. 35 mi. 13.
Super cauda nihil vlterius agetur, sufficiunt enim priora, per quę satis liquet à Sole caudã mutuari, quocirca in posterũ de hac supsedebimus,

Talia ex obseruatione constant.

Latitudo Cometæ ab ecliptica gra. 22 mi. 1.
Verus locus Cometæ in ecliptica gra. 4 mi. 32 ♍.
Declinatio Cometæ ab æquatore 30 gra. 11 mi. Septentrio.
Ascensio recta Cometæ 165 gra. 11 mi.

imperial bedchamber to take their course on the cloudiest of nights. All ancient civilisations saw that the heavens were full of signals and warnings for mortals, in such practical matters as the farming seasons, the tides of the sea, and the matching of the phases of the Moon with the menstruation of women. There was no obvious dividing line between sensible calendar-keeping and foolish predictions about personal affairs.

In their observations of comets, too, the Chinese were vainly trying to make astrology an exact science. Comets were more astounding to our predecessors than black holes are to us; after all, a well-known theorem in relativistic astrophysics assures us (and I quote) that 'black holes have no hair'. And the attention lavished on comets by the ancient Chinese is manifest in this account of troubled times three thousand years ago, in the eleventh century BC:

> King Wu marches on Zhou, faces the east and greets Jupiter, arrives at Qi, which floods, reaches Gongtou, which falls; a comet appears, giving its handle to the people of Yin.

The 'handle' was what we should call the tail of the comet, and the Chinese tried to determine at whom, precisely, the comet directed its threatening semaphore. The astrologers realised that bad news for one kingdom might be good news for someone else.

This narrative comes from *The Book of the Prince of Huai-Nan*, who was Liu An, a taoist prince who lived in the second century BC, much later than the events reported here. Naturalists in his court compiled the book which Needham calls 'one of the most important monuments of ancient Chinese scientific thought'. That its authors should think a comet to be still worth mentioning, nine centuries after the event, speaks of more than the grip of superstition. Astrology begat astronomy, but it cannot escape paternal liability for history and political science too. Why bother to record the mishaps of previous generations, except to try to learn the connections between events in the sky and on Earth? The first archives were astrological, and astrohistory was the foundation of statecraft. Liu An's scholars thought it self-evident that a war and a comet were related; the only problem was to learn how to read aright the ambiguous signal in the sky.

For maoism as for taoism the slogan is, 'Let the past serve the present'. In AD 1978, the planetary astronomer Y. C. Chang was trying out a new computer at the Purple Mountain Observatory and with it he reckoned that the victorious King Wu saw the

Facing: Halley's Comet 1531 described by Petrus Apianus of Ingolstadt. He was the first medieval western astronomer to notice that comet tails point away from the Sun, as illustrated here. Astronomicum Caesareum *(1540) was dedicated to the Emperor Charles V and his brother Ferdinand. (Royal Astronomical Society.)*

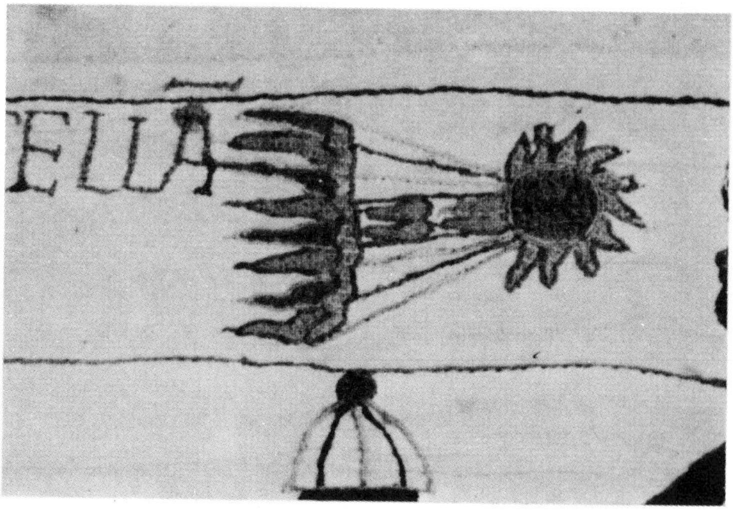

The appearance of Halley's Comet 1066, as depicted in the Bayeux Tapestry, has curiously imagined details, but the overall form is not unrealistic, as shown by comparison with a photograph of Halley's Comet 1910. (Hale Observatories.)

comet *Halley* making one of its regular appearances in 1057 BC. If Chang were right he would fix both a useful date in the early history of his nation and the earliest known visit, by far, of that comet. He also offered other comet sightings in 613 and 466 BC as *Halley*, busily plaguing the human species.

Alas for comet lovers, a leading American authority on *Halley*'s orbit, Donald Yeomans of the Jet Propulsion Laboratory, has questioned these identifications, chiefly on the grounds that Chang's method of computing the orbit back through time did not allow for disturbances of the comet by close approaches to the Earth in AD 374, 607 and 837. These might push the apparition of *Halley* back to 1059 BC and make the historical link doubtful. But the Chinese could not help seeing *Halley*, and if extremely early identifications have not yet been made that is due more to modern problems of computation than to drunkenness among the imperial astronomers. Yeomans is himself engaged, with Tao Kiang of Dunsink in Ireland, in tracing the comet back through time, apparition by apparition, and he regards even Chang's 466 BC identification as 'uncertain'.

A Chinese comet report of 240 BC, often mentioned as the oldest definite sighting of *Halley*, is for Yeomans only 'possible'. The first 'fairly definite' record of the comet, in his judgement, is the one quoted by Chang for August and September 87 BC when, as a book of the Han dynasty, *Outlines of the Universal Mirror*, says tersely, 'a comet appears in the east'. No one can yet be quite sure that *Halley* was on even roughly its present orbit before that,

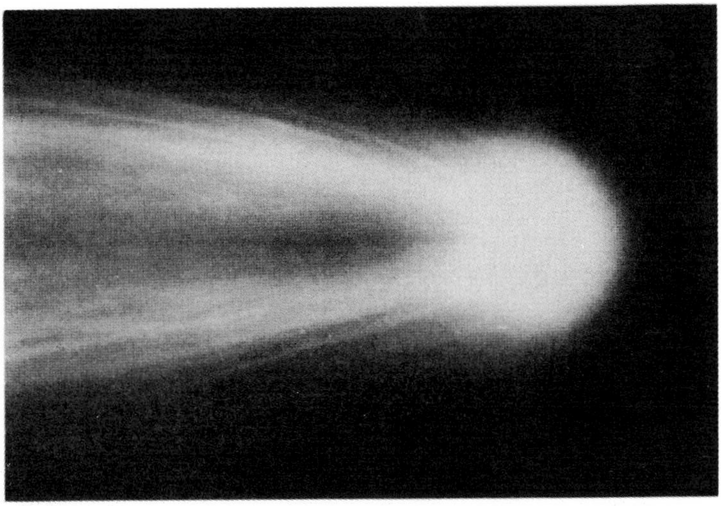

because careful tracking of the comet and the planets is needed, to ensure that there were no perturbing encounters. So suppress any premature urge you may feel to seize on earlier dates and be content that the long-playing comet now heading our way was seen in 87 BC by Julius Caesar at the age of fourteen, while Marius was massacring the Roman aristrocrats with a dedication that foreshadowed Nero's.

Europe can claim the earliest known representation of *Halley* by people who actually saw it, as opposed to drawings made retrospectively to illustrate history-books. The comet put in an appearance in the spring of 1066, before the Norman invasion of England. It was bad news indeed for King Harold, who was to perish that year at Hastings. The apparition's role was sufficiently clear to all concerned for it to be embroidered a few years later, on the seventy-metre linen history of the Conquest, the Bayeux Tapestry. There the comet resembles nothing so much as a spaceship – an eleventh-century prototype of the starship *Enterprise*, complete with motors and missiles. The onlookers' eyes are popping and the caption reads, 'They wonder at the star'. King Harold sits on his throne looking as if someone has just slapped the royal ear with a wet fish. (See colour illustration facing page 16.) The strip cartoon stitched by the eleventh-century ladies remains the most graphic of all representations of comet fever.

Three apparitions later, in 1301, the Florentine painter Giotto di Bondone saw *Halley* and incorporated it realistically in a fresco

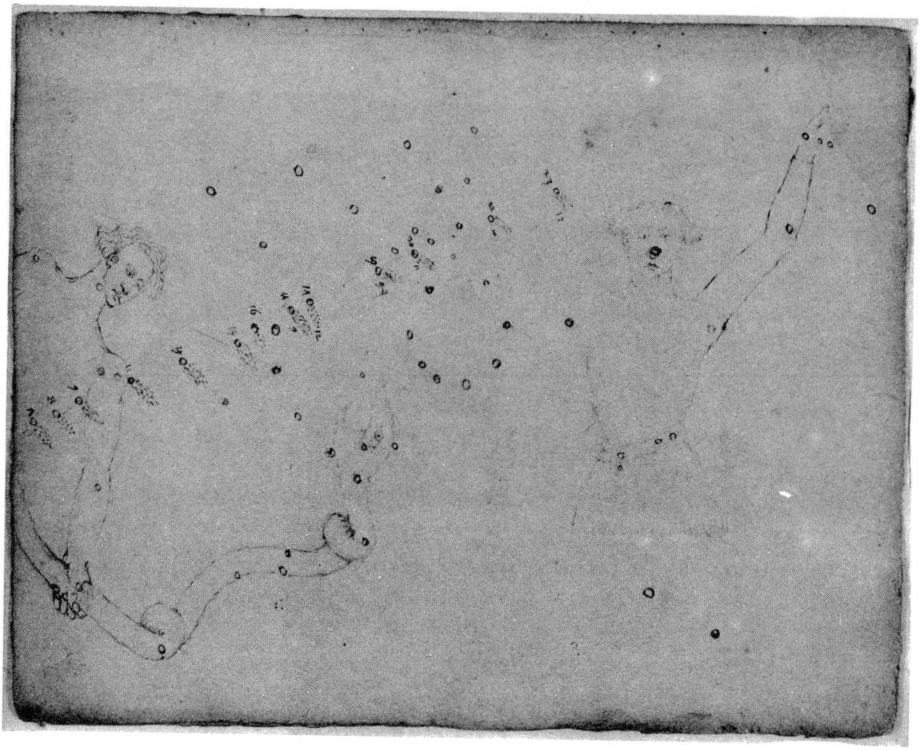

Paolo Toscanelli's plot of Halley's Comet 1456 shows the comet tracking north-west across the sky. The constellation of Ophiuchus is the man on the left wrestling with the Serpent; the loop of stars is Corona Borealis and the man on the right with the luminous nose (the star Nekkar) is Boötes, the supposed inventor of the plough.

at Padua that showed the 'wise men from the East' adoring the infant Jesus. In offering a comet as the Star of Bethlehem that led these astrologers to the birthplace of Christ, Giotto was following a muted tradition going back to the third century AD, that some comets might bear glad tidings. He had no reason to imagine that the comet he witnessed in Italy was the same one as the Magi saw, although another medieval notion was that all comets were just one object coming and going like a whore to confession. When Edmond Halley, four hundred years after Giotto, diagnosed its habits there was a scholarly stampede to prove that the comet *Halley* was the Star of Bethlehem. During the only apparition anywhere near the right date, modern computations have *Halley* rounding the Sun on 5 October in 12 BC. The margin of uncertainty about Christ's birth is not wide enough for that because the bloody deeds of King Herod during the story of the Nativity set limits to the date.

Was it another comet? David Hughes of Sheffield, who is a comet scientist as well as a delver into ancient astronomy, has made out a thorough case for the 'star' being no comet at all, but a conjunction of planets. The appearance close together in the sky of the slow-moving planets Jupiter and Saturn in the constellation of Pisces was a rare event that zoroastrian priests in Babylon would take very seriously. God (Jupiter) with the Jews (Saturn) meant that a righteous Jewish Messiah was about to be born in Palestine (Pisces). Because of the Earth's own motions the planets seemed to come together three times in 7 BC and in *The Star of Bethlehem Mystery* (1979) Hughes neatly relates these three appearances to the phases of the gospel story and agrees with earlier suggestions that the astrologers would have expected the birth of the Messiah to occur on 15 September. Whether Jesus was born that very day is unproven but the year 7 BC, so Hughes concludes, is 'highly probable'.

To rehearse the reports of each of *Halley*'s dozens of returns would be tedious, but the one in AD 1456 was typical. Among the small boys who would have seen it that year were Christopher Columbus and Leonardo da Vinci, while Leonardo's later tutor in mathematics and astronomy, Paolo Toscanelli, was busy plotting the motions of the 'big and terrible' comet. He wrote:

> Its head was round and as large as the eye of an ox, and from it issued a tail fan-shaped like that of a peacock. Its tail was prodigious, for it trailed through a third of the firmament.

Stories that the Pope ordered special prayers for protection against the comet, or even excommunicated it, do not seem well founded, but many Europeans were distressed because the Turks were at the gates of Belgrade and 'mathematicians' were interpreting the comet to mean plague, famine or some other disaster. The very word disaster means 'evil star' and, like 'thank your lucky stars', shows how astrology still invests our language.

Queen Elizabeth I is celebrated in English history for rallying the country against the Spanish Armada, but to her courtiers she showed her mettle long before that. In 1577 a great comet appeared and they tried to prevent the Queen from tempting providence by looking at it. Defiantly she strode to the window and gazed upon the comet, declaring: 'The die is cast.' Two hundred leagues away on a Baltic island, Tycho Brahe was launching comet science by establishing that the comet of 1577 was a long way off in space.

Like other founders of modern astronomy Tycho was himself a believer in astrology, and the early sceptics were outsiders who did not earn their livings that way, notably Thomas More, whose

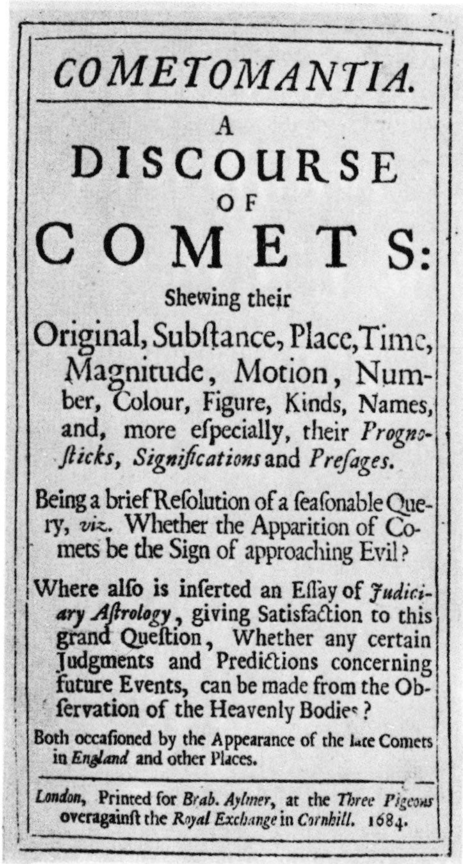

COMETOMANTIA.

A

DISCOURSE

OF

COMETS:

Shewing their

Original, Subſtance, Place, Time, Magnitude, Motion, Number, Colour, Figure, Kinds, Names, and, more eſpecially, their *Prognoſticks, Significations* and *Preſages.*

Being a brief Reſolution of a ſeaſonable Query, *viz.* Whether the Apparition of Comets be the Sign of approaching Evil?

Where alſo is inſerted an Eſſay of *Judiciary Aſtrology*, giving Satisfaction to this grand Queſtion, Whether any certain Judgments and Predictions concerning future Events, can be made from the Obſervation of the Heavenly Bodies?

Both occaſioned by the Appearance of the late Comets in *England* and other Places.

London, Printed for *Brab. Aylmer*, at the *Three Pigeons* overagainſt the *Royal Exchange* in *Cornhill*. 1684.

Title page of Cometomantia, *published in London in 1684 in the wake of 'the late comets' that included that of 1680 and Halley's Comet 1682. (Royal Astronomical Society.)*

Facing: Nineteenth-century scepticism. The caption to Daumier's cartoon of 1858 reads: 'Ah, comets . . . that always means bad luck! No wonder poor Madame Galuchet died suddenly last night!' (From the collection of Marcel Lecomte.)

head was cut off by Elizabeth's father. In *Utopia* (1516) More wrote that his fictional islanders were active in astronomy:

But as for astrology – friendships and quarrels between the planets, fortune-telling by the stars, and all the rest of that humbug – they've never even dreamt of such a thing.

By the seventeenth century Oxford and Cambridge Universities were taking astrology off the syllabus. And in *King Lear* (a play which came out in nice time for an apparition of the comet *Halley* in 1607) Shakespeare has the villain Edmund say: '. . . an admirable evasion of whoremaster man, to lay his goatish disposition on the charge of a star.' But the retreat from astrology was slow and Tycho's discovery was brushed aside, or at least was held not to have proved that all comets were remote. The author of *Cometomantia* (1684), writing in the interregnum between medieval

– Ah!...les comètes ....., ça annonce toujours quelques grands malheurs !..... je n'm'étonne plus que c'te pauv' madame Galuchet est morte subitement hier soir !...

and modern astronomy, put the following gloss on the well-known vulnerability of kings to comets:

If it once be admitted that comets distemper and inflame the air, and exhaust the *succus* [i.e. juices] of the Earth, it will necessarily follow, that a barren soil, and the corrupting and blasting of the fruits must be the products of them: and from these will naturally ensue dearth, scarcity and famine. And, as the inevitable effect of both, we must expect sickness, diseases, mortality, and more especially the sudden death of many Great Ones, because these are sooner and more easily hurt than others, for their delicate feeding, and luxurious course of life, and sometimes their great cares and watchings, which weaken and infeeble their bodies, render them more obnoxious than the vulgar sort of people.

Relish the old and literal use of 'obnoxious' and give credit to a proto-scientific mind for seeking a chain of cause and effect that might make sense of inherited beliefs.

The scrupulous if misguided magic, that brought the magicians to Bethlehem and moved the Chinese to invent mechanical clocks, was doomed to dilapidation. As confidence tricksters found, you can say anything with a straight face and someone will believe it; Casanova, for instance, used astrology to ingratiate himself with women. The art is, of course, illegal in England, where for a century and a half the law has declared:

> . . . every person pretending or professing to tell fortunes or using any subtle craft means or device, by palmistry or otherwise, to deceive and impose on any of His Majesty's subjects . . . shall be deemed a Rogue and Vagabond, within the true intent and meaning of this Act.

The lawmakers of King George IV's time prescribed a penalty of three months' hard labour, but the act is flouted in the astrological columns of newspapers and magazines, where Rogues and Vagabonds deceive Her Majesty's subjects every day of the week.

Yet the greatest imposition on the populace continues to come from official fortune-tellers. Two millennia after Nero, and despite repeated disappointments, a dogged belief persists: the future well-being of nations and the behaviour of their peoples is divinable, and if policies conform to what the signs portend everything will be all right. The governments' astronomers are left in peace nowadays to look for black holes, and tongues might wag if the leaders of modern industrial nations were seen to peer at the sky when drawing up their budgets. No need for that: the most respected of latter-day astrologers have adapted to the times, exchanging their magic robes for city suits and their divining boards for computer print-outs. Still meddling in statecraft, they call themselves economists.

Traditional astrology remains a world-wide multi-million-dollar industry and lunar astrology is making a big play in the 1980s, exploiting the correspondence with the emotional rhythms of the menstrual cycle. When the National Academy of Sciences in Washington commissioned its centenary statue of Einstein the sculptor elected to show the old spinozan sceptic gazing at his horoscope – a prospect so grotesque that the statue was seriously undersubscribed by the scientific community, even when the sky map was amended. But, to the more gullible, comets still telegraph disastrous news; in 1973, for instance, a pamphlet of the Children of God explained that the comet *Kohoutek* heralded the end of the world on 31 January 1974.

And just a lifetime ago, when astronomers foresaw that on 18

May 1910 the Earth would pass through the tail of the comet *Halley*, there was consternation of a different, semi-scientific sort. The encounter was to occur far from the head of the comet and experts knew how thin a comet tail really was; also that the Earth had passed unscathed through at least two comets' tails in the nineteenth century. But their detection of cyanogen gas, a known poison, in the tail of the comet *Morehouse* in 1908 more than outweighed their reassurances. And a French writer suggested that hydrogen in the tail might react with the Earth's

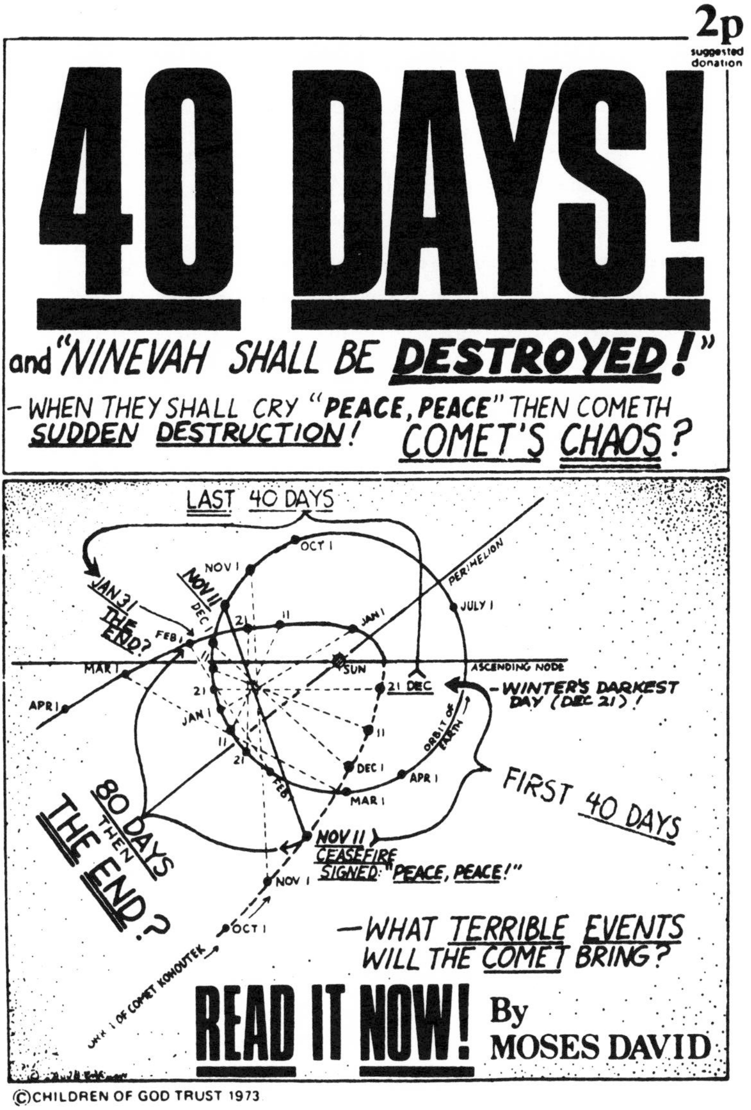

Twentieth-century credulity. That comets are still good for a prophecy of doom, appears in this pamphlet of 1973.

atmosphere, 'overwhelming our planet in a gigantic explosion'.

The man and woman in the street awaited the event with apprehension. Inevitable claims that the 'official' calculations were wrong, and *Halley*'s head would hit the Earth, added to the worries. While some people sealed their windows against the cyanogenous fall-out there were others who, as mentioned earlier, saved the comet the trouble and killed themselves. The more daring kept watch as the Earth swam into *Halley*'s tail.

One social historian who noted the public mood was James Thurber, sixteen years old at the time. He was aware of certain predictions that *Halley* was going to strike the planet somewhere between Boston, Massachusetts, and Boise, Idaho, and knock it into the outer darkness, far from the Sun. As Thurber recorded:

> Nothing happened, except that I was left with a curious twitching of my left ear after sundown and a tendency to break into a dog-trot at the striking of a match or the flashing of a lantern.

The Chinese Emperors were right: the less the general public hears or sees of comets, the better. A recent remission in the fever owed more to electric street lighting, which made comets hard to see, than to wisdom in cosmic matters. Revelations about the trashy nature of comets and their obedience to the law of gravity failed to eradicate comet fever, which remains endemic because nature feeds a human appetite for cheap thrills by tossing in a great comet every ten years or so. The chief blame falls squarely on one comet and one man.

2.
SHARPLY VEERING WAYS

Halley would count as a famous man of science had he not attatched his name for ever to a cosmic bauble. Everyone hears of Parkinson's Disease while James Parkinson is forgotten, and Lord Cardigan is immortalised for his choice of knitwear, not his equally original misadventure in battle. And Edmond Halley the natural philosopher, friend and pest of Isaac Newton, is outshone by Halley's Comet. He meant to put comets in their rightful place in the cosmic scheme and thereby allay all speculations and fears about them, but in the end he made the fever worse. It was his misfortune to be born in 1656, when comet studies were coming to a crisis and 'his' comet was just twenty-six years away, heading sunwards.

Otherwise everything was in order for Halley to go down in history as a run-of-the-mill genius and a personable fellow. He had a suitably prosperous yet obscure father, a soapmaker of London. As a young man he demonstrated his talents as an astronomer by making the first accurate charts of the southern sky and, later in life, he began the study of the private lives of the stars. Halley could be profound, as in his remark that if the universe were an infinity of stars, the whole sky ought to be very bright. He was imaginative, and thought that there were people living inside the Earth. He was brave, too, for he invented a diving bell and tested it himself, and took command of a small warship, *Paramore*, for two voyages into the South Atlantic to plot the global variations in the Earth's magnetism.

In the same ship Halley carried out the first thorough survey of the tidal currents of the English Channel. He climbed a mountain with a barometer to find out how the pressure of the air diminishes with height, and he mapped the winds of the world, including the trade winds and monsoons. His originality appeared in little things, for instance measuring the areas of English counties by cutting up the map and weighing the pieces, as much as in notable projects that included drawing up tables of life expectancy. In these various guises Halley can be counted the founder of modern cosmology, geophysics, oceanography, meteorology and demography, as well as stellar astronomy. For all this he deserves a respectful remembrance.

But Paris is a city of snares for many a young man and it was there, at the age of twenty-four, that Halley became infatuated with comets. His visit coincided with the appearance of the great comet of 1680. He obtained full records of its movements across the sky, from the director of the Paris Observatory, Giovanni Cassini, and attempted to plot its route through space. He made a thorough mess of the task because he thought that comets travelled in straight lines. Cassini, for his part, supposed that the comet was on a small orbit around the Sun, and that was equally false.

The ideas of early scientists often seem crass because science is a garden where theories grow in a bed of facts, and errors are eventually weeded out; we have acres of facts that were quite uncultivated three hundred years ago and, as a result, any third-class graduate of the twentieth century knows far more about the workings of nature than, say, Isaac Newton did. But the notion that comets went in straight lines was advanced thinking in Halley's time, when the attempts to deal factually with comets and their orbits were proving just as muddy as the fictions of astrology. So let us leave the young Englishman scribbling, crossing out his figures and mopping his brow, as he tries to impress the French *virtuosi* with his analytical powers, and see how he comes to be making such an ass of himself.

If an astronomer sees a light slowly shifting its position in the sky from night to night, he cannot tell intuitively which way it is travelling because he does not know how its distance is changing. Watchkeepers at sea are similarly puzzled to know whether a light is *Queen Elizabeth 2* ten sea miles away or a yacht a hundred metres off, but the worst that can happen to the sailor who misjudges lights at sea is death by drowning. The seventeenth-century astronomer who came up with the wrong answers about lights in the sky might find himself in Hell.

Some lights were not supposed to be there at all and the 'new' stars that Chinese astrologers recorded over the centuries passed unnoticed by their European counterparts. The wiseacre who unwittingly blinded western civilisation for two millennia was the Macedonian called Aristotle. He performed at the Lyceum in

Edmond Halley

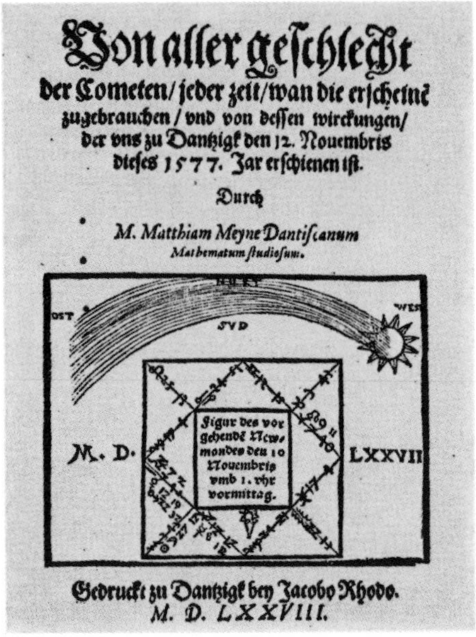

The famous comet of 1577 examined in an astrological broadsheet from Danzig.

Facing: Seventeenth-century puzzlement about the orbits of comets, captured on the title page of the tome of Hevelius, 1668. The author is showing his version in the centre, but it is quite wrong. (Royal Astronomical Society.)

Athens in the fourth century BC and told anyone who cared to listen how the universe was divided into two distinct realms: the 'sublunary' regions below the Moon, full of change, decay and comets, and the farther regions beyond the Moon where, among planets and stars, quite different rules applied and nothing ever changed.

This blossom of the ancient imagination should have been short-lived and harmless, except that it appealed to the medieval church which adopted the perfect unchanging cosmos as dogma. The theory was self-protecting. In the first place Aristotle's teacher, Plato, had decreed that humour had no place in philosophy, so you weren't allowed to laugh off such propositions. The scheme of the universe also made astronomy very labour-saving: why bother to look at the fixed stars if nothing ever changed? And if you didn't look, you wouldn't see that Aristotle was wrong. Nature eventually put matters right with a change in the realm of the stars that was hard to miss.

One evening in 1572 surprised peasants were importuned by a young Danish astrologer with a golden nose. Tycho Brahe, who had been partly defaced in a German duel, was pointing in agitation at the sky and asking the peasants what they saw. The astrologer did not believe his own eyes, because what he spied while walking home from his laboratory at Heridsvad Abbey was

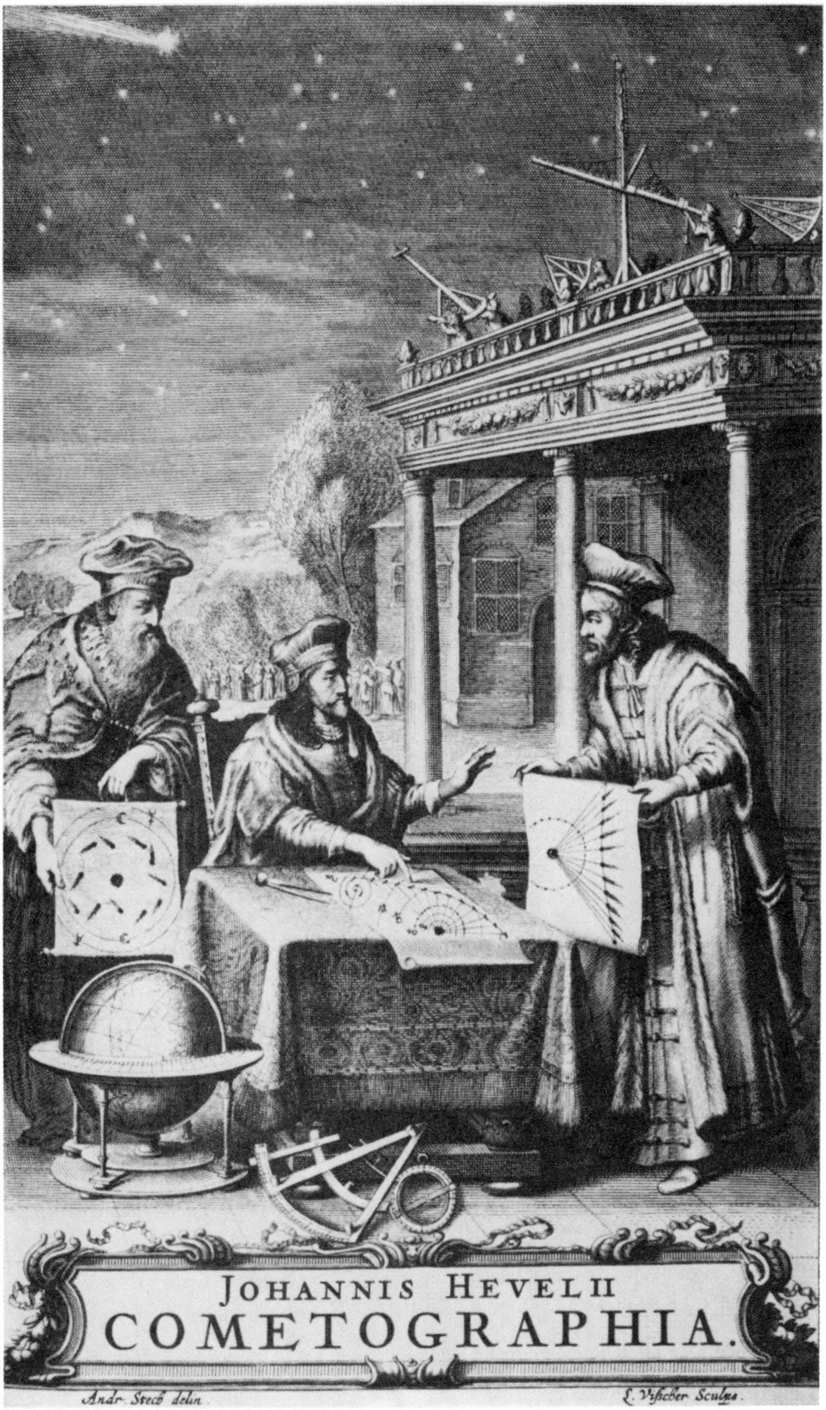
JOHANNIS HEVELII
COMETOGRAPHIA.
Andr. Stech delin.
I. Visscher Sculps.

in sudden contravention of Aristotle's laws of the universe, laid down 1900 years earlier. But the peasants, their eyesight unimpaired by learning, confirmed that they could see a very bright star that they had never noticed before, in the breast of Cassiopeia. In truth it was another of those exploding stars but that was not understood until the twentieth century; in any case the cause was less significant at the time than the bare fact of a 'new' star.

It is hard to grasp the horror and incredulity of that night four hundred years ago. Tycho wrote afterwards:

> A miracle indeed, either the greatest of all that have occurred in the whole range of nature since the beginning of the world, or one certainly to be classed with those attested by Holy Oracles. . . .

Scholars of Europe who could bring themselves even to see the intruder wished to say that it was a sublunary event close to the Earth, 'like a comet'. But Tycho measured the angles between the new star and other stars in the sky and showed that they did not change through the night. Therefore it was not a sublunary apparition but very remote. After six months' study – by which time the new star had faded – Tycho was satisfied that it could not even be a planet but lay 'in the eighth sphere, among the other fixed stars'.

Lest anyone should feel at all smug about this incident, let me say that modern science has been through a similar trauma about a light in the sky and how far away it might be. By the mid-twentieth century, astronomers were self-satisfied because they knew how the universe generated light, by nuclear reactions in stars. Then one day in 1963 the young Dutch astronomer Maarten Schmidt, working in Pasadena, was puzzling over unidentified features in the light of a bright blue quasar, known to be a strong source of radio waves. He solved the problem but the answer was incredible and Schmidt, too, mistrusted his eyes. He had to ask an American colleague to confirm that what he was seeing in the photographs really meant that the quasar was an immense distance away. If so, it was far more powerful than the most violent nuclear explosions could explain. When Maarten Schmidt described his feelings it could have been Tycho Brahe speaking: 'That night I went home in a state of disbelief. I said to my wife, "It's horrible, something terrible happened today."' The quasar needed a new source of energy to explain it and nowadays such objects are thought to contain massive black holes that are gobbling up entire stars.

Facing: A volcanic eruption far off in the Solar System is reminiscent of an origin for comets – almost certainly false – with which astronomers have toyed for centuries. This explosion on Io, a moon of the planet Jupiter, was fortuitously recorded by a passing American spacecraft, Voyager 1, on 4 March 1979; the plume is about 150 kilometres high. (Jet Propulsion Laboratory.)

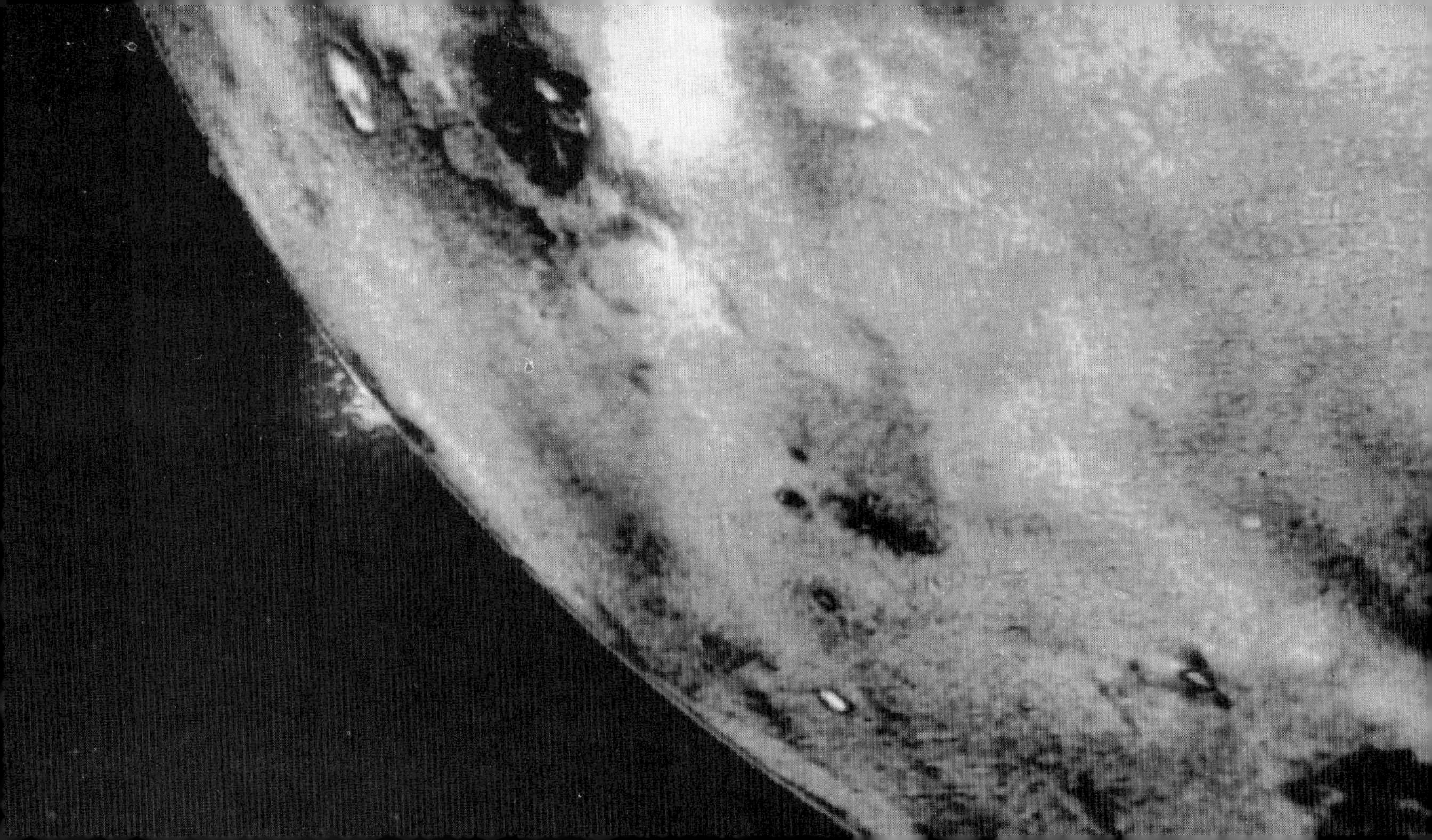

The King of Denmark gave Tycho an island and funds to build an observatory. This island was Hven, lying in the Sound between Denmark and Sweden, and here Tycho made a special study of the great comet of 1577. The result was almost as mind-boggling as the new star. For all his astrological showmanship (he used to dress up in fancy robes to look at the sky) Tycho had shrewd scientific instincts. If Aristotle was wrong about the stars, perhaps he also erred in supposing that comets were phenomena of the upper air, like the aurorae of the polar skies. Tycho remembered an Arabic suggestion that 'comets are produced not in the air but in the heavens' and in his account of the new star he had declared his plan: 'That this can be the case is not yet clear to me. But please God, sometime, if a comet shows itself in our time, I will investigate the truth of the matter.' He could not have asked for a more suitable example than the comet of 1577, bright enough to be visible in daylight.

As soon as someone as skilled as Tycho took the trouble to check Aristotle's fable about comets being sublunary exhalations the observations falsified it. This time there was nothing to be gained by simply observing the position of the comet from one place (it was obvious that comets moved in relation to the stars) but Tycho compared the apparent positions of the comet as recorded at Hven with simultaneous observations made in other parts of Europe. At separations of a few hundred leagues the direction of the Moon was measurably different from two observatories, but no perceptible differences appeared in the direction of the comet, which meant it was far beyond the Moon. More pious astronomers disagreed, but Tycho collated the observations too thoroughly to be wrong.

Tycho's island belongs nowadays to Sweden and nothing remains of the palace-observatory of Uraniborg, the first European venture in Big Science. Tycho was a tyrant who imprisoned tenants when they were in arrears with their rent and, on a visit there by boat, I imagined the satisfaction with which the islanders looted the stones, when even the Danish royal court could put up with the proud and extravagant astrologer no longer and Tycho carted his instruments away to foreign lands. After the islanders came the collectors for museums and only a few molehills remain, which once housed underground instruments. There is a statue of the hated landlord, now apparently forgiven, perhaps because Tycho's study of the 1577 comet helped to undermine the divine rights of all kings and landlords.

Facing: Donati's Comet as seen over Paris in October 1858. The comet and the artist have conspired to show, almost diagrammatically, two types of comet tails: the broad, curved dust tail and the narrower and straighter plasma tails, of which two are visible here. Giovanni Donati first spotted the comet four months earlier.

(This picture shown in color on front cover.)

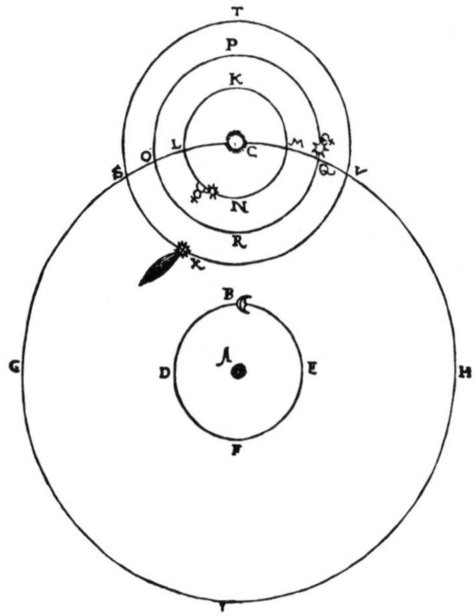

Trying to make sense of comet tracks – 1. Tycho Brahe supposed that the comet of 1577 was in orbit around the Sun, but the Sun was in orbit around the Earth. This peculiar compromise between the ancient and copernican views of the Solar System found little favour.

In the medieval scheme of the universe, feudal notions of rank in church and state were advertised in the sky, where the Sun was King and the planets were lesser lights, somewhat unruly and potentially mutinous. Angels and barons, eagles and flies, trees and herbs, precious stones and common soil, all had their places in the hierarchy under God, and angels of descending degree had charge of the fixed stars and the planets. More technically, the Earth was at the centre of the universe, and the Moon, the Sun and the five known planets were perfect bodies, each fixed on a perfect crystal sphere that revolved about its axis at the appropriate rate. The spheres nested inside one another like Russian dolls and the eighth sphere, mentioned by Tycho, carried all of the 'fixed' stars. To doubt any of this was a grave matter. William Shakespeare has Ulysses explaining it:

> The heavens themselves, the planets, and this centre
> Observe degree, priority and place,
> Insisture, course, proportion, season, form,
> Office and custom, in all line of order . . .

And so on. In 1600, just about the time when Shakespeare wrote *Troilus and Cressida,* Giordano Bruno was burnt at the stake in Rome for his heretical views, including the opinion that the Earth, 'this centre', moved.

Galileo Galilei was soon seeing unseemly sights through his

telescope: sunspots, craters on the Moon, the satellites of Jupiter and so on. He found himself up before the Inquisition too, for his opinions on celestial matters. In 1633 he had to recant and confess his error and, even so, was condemned to internal exile. Not that Galileo, dying lonely and blind, had any reason to be distressed, because his name would be cleared within a few centuries of his death. In 1979 the Pope, on the occasion of Albert Einstein's centenary, declared that the verdict of the Church on Galileo ought to be re-examined, although a Vatican lawyer afterwards cautioned that progress in the case might not be rapid because the protagonists were dead.

The clash between Galileo and the Church was no schoolroom quibble about the right way to analyse cosmic motions: the new astronomy, and more particularly its dissemination to students and the public, threatened the hierarchies of Heaven and Earth. In Protestant England at the end of the seventeenth century, Halley and his friends had no fear for their skins, but jobs were another matter and Halley's appointment to a professorship at Oxford was blocked for some years because he was 'a banterer of religion'. Thus seemingly boring and technical questions about the tracks of small pieces of the universe had to be handled circumspectly, because they could get the analyst into trouble with God, or at least with his representatives on Earth. For a hundred years comets had a certain philosophical importance, as the hand-grenades of a cosmic revolution.

The far-flung comet of 1577 smashed through the transparent crystal spheres that supposedly carried the planets. For Tycho's later collaborator, Johann Kepler, that simple statement was enough, but Tycho's own version was more complicated. He rejected the radical idea of Copernicus that the Earth moved, but he set the planets and the comet of 1577 in orbit around the Sun. That put the sphere of the Sun and the spheres of the planets into a fatal tangle. And when Kepler worked over Tycho's careful measurements of the motions of the planets themselves he found that they did not even travel in circles. If the crystal spheres still existed after that they would have to be very wobbly.

Kepler, the German exile who became royal mathematician, magician, and musician of the planets at Prague, inaugurated modern planetary astronomy by discovering that the orbit of a planet around the Sun was always an ellipse, the somewhat squashed circle that you get if you slice obliquely through a cone. He also thought that the speeds of the planets made a musical scale. Kepler's magic of conic geometry and planetary tunes was an attempt to puzzle out the intentions of a mathematically-minded Creator. That has been the policy, at least metaphorically, of astronomers and physicists ever since. And Kepler thought that God must cause an intruding comet to proceed through the heavens in a straight line.

To seek the correct track for a comet was like pondering the right shape for wagon wheels – a question to which Galileo turned his attention in his days of freedom. He put his considerable authority behind the opinion that triangular wagon wheels would be less perfect than round ones, so lampooning the scholars who wanted to rank the geometric figures according to pedigree and nobility. But Galileo's plea for a more democratic and functional view of shapes was scarcely heard above the noise of the cosmic machinery breaking down.

The suitability of circular orbits for the eternal motions of the planets was so self-evident that medieval minds had clung to them in the face of any observations to the contrary. A perfect planet mounted on a perfect crystalline sphere naturally revolved for ever and the angels were no more than casual machine-minders for an engine endowed with perpetual motion by virtue of its shape. In contrast earthly objects, going in straight lines, quickly

COMETA ORIEN-
TALIS,
Kurtze vnd eygentliche Beschreibung deß newen Cometen/so im November deß abgelauffenen 1618. Jahrs in Orient/oder gegen Auffgang der Sonnen allhie erschienen/vnd von menniglich gesehen worden.
Auß warhafften Astrologischen vnd Historischen Gründen menniglich zur Nachrichtung gestellt/
Durch
M. Gothardum Arthusium Dantiscanum, P.C.Histor. & Philo-Mathem. zu Franckfurt am Mayn.
2. Timoth. 3.
Instabunt tempora periculosa.
Franckfurt am Mayn/bey Sigismundo Latomo.
M. DC. XIX.

Facing: Trying to make sense of comet tracks – 2. Johann Hevelius thought that comets went roughly in the straight lines proposed by Johann Kepler, but were somewhat deflected. Notice how the comets are supposed to spiral out of planets and orientate their 'disks' to face the Sun.

The comet of 1618, here portrayed in a publication from Frankfurt, was one of those that Johann Kepler insisted, quite wrongly, travelled in a straight line through space.

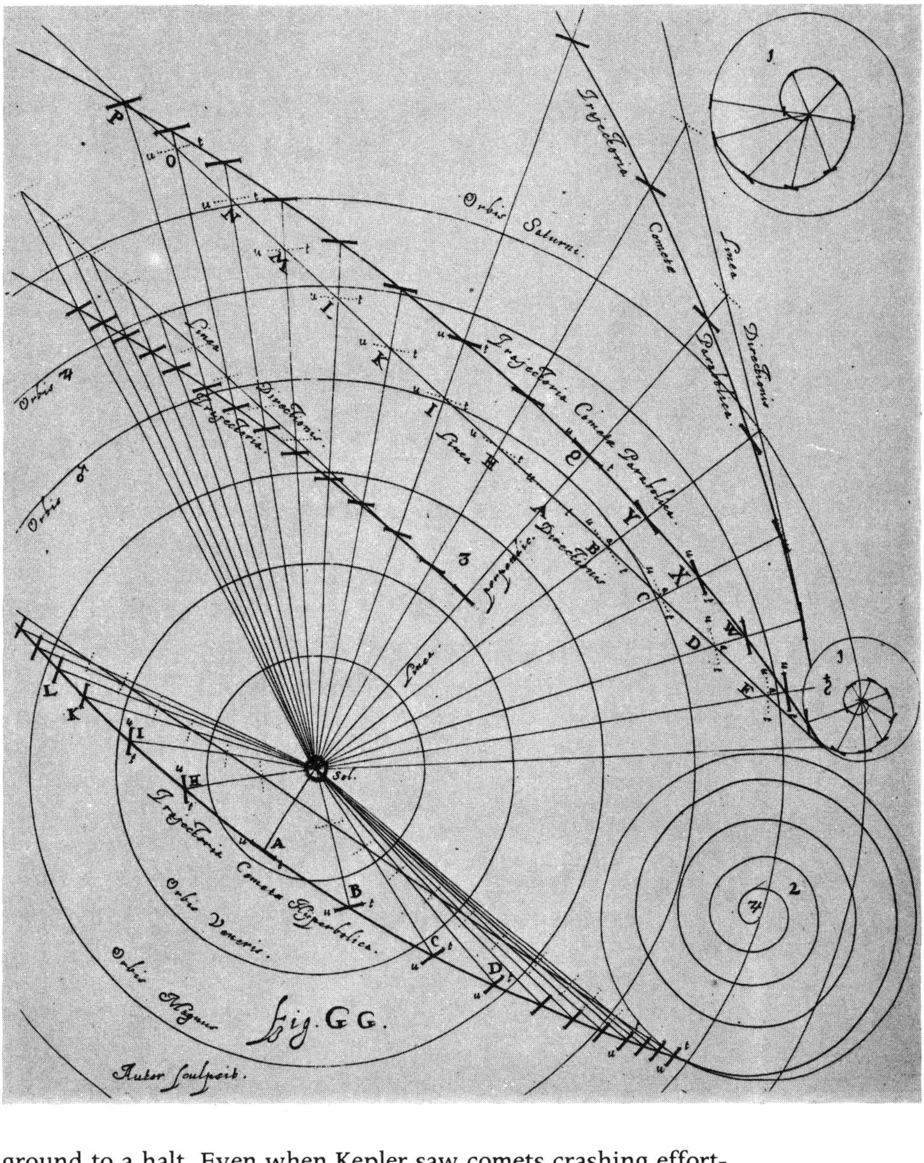

ground to a halt. Even when Kepler saw comets crashing effortlessly through the spheres and he had squashed the remaining circles into ellipses, he could not shake off the ancient habits of thought.

Tycho had put the comets on shortlived segments of circular orbits to give them spontaneous motion, but Kepler frowned on the implication that the ephemeral comets shared any of the divine order of the planets. Recent historical inquiries by James

Ruffner of Detroit clarify Kepler's thoughts: his chief analogy for a comet was a rocket, which ignites, accelerates and then slows down, while travelling roughly in a straight line. A straight line was mortal because in a finite universe it could not be infinitely long. That made it a fitting track for a mortal comet.

After decidedly sloppy processing of the observations, Kepler asserted that the comets of 1607 and 1618 fitted his theory. If the tracks did not look anything like straight lines, Kepler explained, that was because the Earth was moving, thus proving Copernicus correct in saying that our own planet orbits around the Sun. By this argument Kepler put Halley and other successors in a bind; they might seem to be siding with anti-Copernican fuddyduddies if they questioned the straight-line motion of comets. Pierre Gassendi stretched the specious line of reasoning to say that even apparent changes in a comet's speed were illusions due to the Earth's own motions: comets, he declared, travelled eternally at a steady speed in a straight line, across an infinite universe. Applied to comets it was silly, yet it was a possible kind of motion and the galaxies flying apart in the expanding universe approximate to Gassendi's prescription for the flight of comets.

The most elaborate adaptation of the straight-line theory of comets appears in the tome *Cometographia* by Johann Hevelius, published in 1668. That eminent astronomer of Danzig explained that comets were disk-shaped bodies flung out of planets. They hurtled across the Solar System in lines that were 'never exquisitely straight as Kepler and others would have it'. The observations showed they were somewhat curved towards the Sun, and Hevelius devised a theory in which the changing orientation of the disk modified the comet's motion. It was all subtle, complicated and wrong.

This brings us back to Halley in Paris, floundering among his diagrams as he tried to make the comet of 1680 go in a straight line, as the great Kepler said it should. If he suspected for a moment that the suave subjects of Louis the Sun King were laughing up their cuffs, the normally sprightly Englishman must have been abashed. For whatever reason, Halley nursed the problem of cometary orbits like an unhealed wound, until a quarter of a century later he made the prediction that overrode all his more serious accomplishments.

From his domestic observatory at Islington near London, at half past six in the morning of 22 November 1682, when a newly married man might have been better occupied, Halley observed the comet that became his *Doppelgänger*. From that moment events took a course as remorseless as the flight of the comet. A year later, at the Royal Society of London, Halley and Robert Hooke talked about a possible law of gravity that governed the motions of all objects in the Solar System. Christopher Wren offered a small prize to the one who should prove it first, but neither of them could do the sums. In August 1684 Halley had the idea of consulting the brilliant Isaac Newton at Cambridge.

Halley was in his late twenties, quite tall, lean-faced, and affable to everyone. At this fateful meeting, he was full of respect for Newton, who was then in his forties and needed to be handled with tact. In contrast with Halley's high spirits, Newton was a shy, sulky man wrapped up in his studies, slow to publish his results yet jealous of competition. He told Halley that he had solved the problem of universal gravity, although he had mislaid his calculations.

This was as vexing as if Columbus had said he had discovered new lands and forgotten the way, but Halley persuaded Newton to set his thoughts in order and explain to all the world the motions of the Moon, the planets, and the comets too. Newton was interested in a more lucrative art than astronomy, and he had to be nagged into completing the work. The feelings of the two men about this enterprise are best expressed by the fact that Halley, who could not really afford it, paid for the publication of Newton's famous *Principia* in 1687, while the author, who was not hard up, made no contribution to the expenses.

If comet fever were curable at all, Newton's theory might have eradicated it for ever. As Halley wrote in his dedicatory Ode to Newton in the *Principia* (translated from the Latin by Leon Richardson):

> . . . Now we know
> The sharply veering ways of comets, once
> A source of dread, nor longer do we quail
> Beneath appearances of bearded stars.

Mark the 'sharply veering', with which Halley bids farewell to the accursed straight lines of Kepler.

The comet of 1680, which caused Halley so much trouble in Paris, became the prime example of the theory of gravity. Newton had observed it for himself and, with the sure touch of supergenius, he realised that if the comet seen climbing away from the

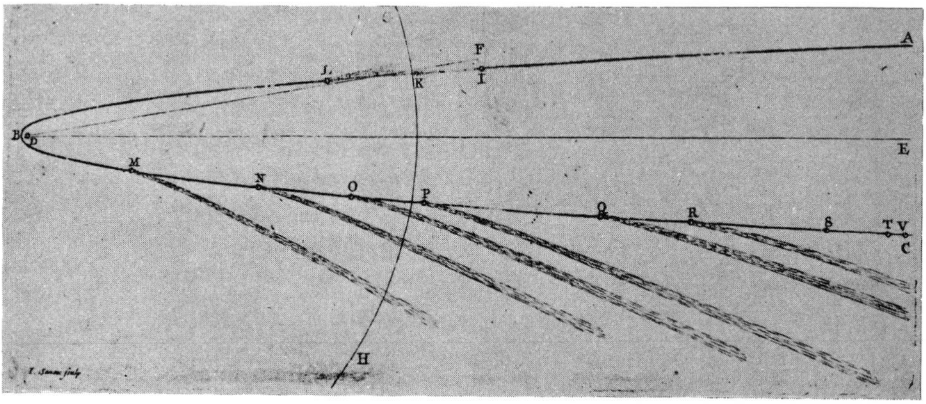

Sun in mid-December 1680 was the same comet as the one seen approaching the Sun in almost the opposite direction in November, it must have changed course. The correct track of the comet, so Newton told Halley, was *roughly* a parabola. He drew a picture of it but left the obliging Edmond to figure out the arithmetic. Neither of them knew that in Saxony Georg Dörffel had already arrived at that very answer for the comet of 1680, but he was not drinking from the same bottle.

For Newton comets were the exception that proved the rule of his all-embracing law of the gravitational force. He explained the elliptical orbits of the planets, inferred by Kepler, as an effect of the action of the Sun's gravity, diminishing with distance. But any object passing through the Solar System must be subject to the same gravity, and under that law there were only five possible tracks: a suicidal line straight into the Sun, a circle, an ellipse, a parabola or a hyperbola. The last two, like ellipses, are 'conic sections', shapes that can be obtained by making different slices through a cone. Unlike ellipses they do not catch their own tail but begin and end infinitely far off – a hyperbola would be the track of an object making a one-off encounter with the Solar System, rushing through and away, never to be seen again. A parabola, on the other hand, is an ellipse stretched to infinity and, even while he showed that the comet of 1680 fitted a parabola, Newton really thought of comets travelling on very elongated ellipses, scarcely distinguishable from parabolas: 'I am out in my judgement, if they are not planets of a sort, revolving in orbits that return into themselves with a continual motion.'

Instead of being satisfied with this general answer to his problem, and leaving well alone, Halley continued to embroil New-

Facing: Trying to make sense of comet tracks – 3. Isaac Newton correctly inferred that the comet of 1680 went in a 'sharply veering' orbit around the Sun. It approximated to a parabola but Newton suspected, again correctly, that a typical comet went in a very elongated ellipse, almost indistinguishable from a parabola.

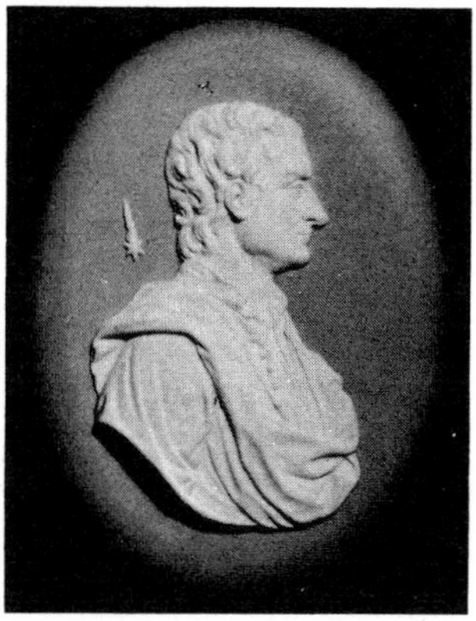

Isaac Newton commemorated on a Wedgwood medallion of the late eighteenth century. It is complete with Halley's Comet, the reappearance of which in 1758–9 was regarded as a vindication of Newton's theory.

ton in the detailed study of comets, and the second edition of the *Principia* published in 1713 contains an extraordinary number of pages devoted to the subject. The most tiresome passage lists every recorded observation of the 1680 comet, whether from Venice, the East Indies or from 'Maryland, in the confines of Virginia', where one Arthur Storer saw the comet close above the star Spica at five o'clock one morning, by the banks of the Patuxent, and so won himself a mention in the grandest book in science.

Meanwhile Newton became as mad as a hatter: by 1692 he was suffering from depression, paranoia, insomnia and forgetfulness, and his hands shook. Poor Newton's scientific work was impaired but in that state he was judged fit for public office and went on to become Master of the Mint and a Member of Parliament.

Newton had weightier matters than comets disordering his mind: he was into alchemy and trying so passionately to make gold that he often slept in his laboratory. The mercury (the liquid metal, not the planet) that figured in his experiments evidently poisoned Newton, along with lead, arsenic and antimony that he also used. When makers of beaver hats treated their furs with mercury they too suffered poisoning of the nervous system – hence Lewis Carroll's Mad Hatter. The environs of Newton's rooms at Trinity College must have been like a mini-Minamata and there are no grounds for blaming Halley's importunities for his friend's affliction.

If Halley's own fine brain was clouded in those later years in orbit around Newton, it was because of the brandy. He acquired a taste for it during his brief naval career, 1698–1701, when he also learned to swear like a sea dog. I seem to hear grouchy Newton and breezy Halley at one of their conferences on comets, babbling like Caliban and Stephano in *The Tempest*, with Newton muddling his planets and his metals and Halley offering to keelhaul every d—d orbit in sight. Be that as it may (and it may not, for anyone with a scruple of respect for the founders of our science), their joint treatment of comets was less than perfect.

Newton's explanation of comets had started Halley on the game that occupied him, on and off, for many years: to find, in the historical records, objects coming back again and again on the same sort of tracks, at long but equal intervals. Halley kept botching it, and his efforts might easily have sunk into a decent oblivion. In one attempt, recorded in the *Principia*, Halley identified the comet of 1680 with three earlier comets seen at intervals of 575 years: in 44 BC (after Julius Caesar died), in AD 531, in 1106 and then 1680, with a return expected in AD 2255. Despite Newton's endorsement, the inference was incorrect according to later reckonings, and the same must be said for Halley's attempt to link a comet of 1661 with one seen in 1532, to imply a return in 1790.

Their heroes' mistakes are often passed over by the chroniclers of science, who thereby hide the human face of science and make discovery seem automatic and easier than it really is, in ways discouraging to non-geniuses. It was not plain sailing to characterise the trajectories of comets long-since passed, by how close they went to the Sun and how their tracks were oriented in space. At the cost of great labour, Halley processed two dozen comets in this way, looking for similarities. And, alongside his two duds, one of Halley's comets was correctly judged. By 1695 he was becoming convinced that the comet he himself witnessed at Islington in 1682 corresponded very well with others seen in 1531 and 1607.

All three comets travelled the 'wrong' way around the Sun, contrary to the direction in which all the planets swirl. The orientations of the orbits agreed in detail and a discrepancy in the periods between the apparitions could be accounted for by the disturbance of the comets by the planets, especially by Jupiter. Therefore the three comets were one, travelling around a highly elliptical orbit that brought it back to the vicinity of the Earth and the Sun every seventy-five or seventy-six years. To clinch the argument, a fourth comet had been seen going the wrong way in 1456, which made it plausibly an appearance of the self-same object.

In 1705 Halley wrote: 'Whence I would venture confidently to predict its return, namely in the year 1758.' At a stroke Halley undid the benign effect of Newton's mathematics, which should have reduced comets to their proper status as lowly curiosities. Perhaps a truculent lieutenant in *Paramore* whom Halley had court-martialled was right, and he wasn't fit to manage a long-boat, never mind one of His Majesty's pinks or the sanitation of the Solar System. At any rate, nothing was more surely calculated to stir everything up again than a plausible forecast of the return of a bright comet within the lifetimes of Halley's younger readers. Newton did not help matters either: as we shall see, he regarded comets as divine consignments of fuel and water, sent to keep the Sun alight and save the Earth from drying up.

Halley and Newton remained on good terms until just before Newton's death, in 1727. Newton was President of the Royal Society and Halley was by then the Astronomer Royal. He was refusing to publish his observations of the Moon, which Newton rightly thought should be public property. It was an inversion of their roles in the matter of the *Principia* and the quarrel was bitter enough for friends to say it shortened Newton's life. But Newton was already in his mid-eighties, and Halley himself was to live to the same ripe old age. The Astronomer Royal died in his chair at Greenwich in 1742, immediately after taking a last swig of alcohol.

In a late note about the comet, published after Halley's death, his mask of bonhomie slipped for a moment to reveal again the embarrassed young visitor to Paris:

> . . . if according to what we have already said it should return again about the year 1758, candid posterity will not refuse to acknowledge that this was first discovered by an Englishman.

His wish was fulfilled with a vengeance. With all Europe impatient for it, the comet was seen again on Christmas Day 1758 by an amateur astronomer, a farmer living near Dresden. Unlike the great mathematician Archimedes, who became known by candid posterity as the Eureka streaker because of a momentary fit of absent-mindedness, Halley had worked for many years to give his name away. His successors made his cometary reputation and marred the rest, by acclaiming the visitor as Halley's Comet.

After 1758 it turned up in 1835 and 1910, and like a yoyo it now spins back again in the 1980s. The name applies retrospectively, so that the comet of 1607, which Kepler swore went in a straight line, was *Halley* up to its usual elliptical tricks, and so were at least twenty-two previous apparitions, back to 87 BC. The intervals between them varied between seventy-five and seventy-nine years, with seventy-six years as the average since 1066.

How Halley pronounced his name is not uncontroversial and possibly of some momentary importance to newscasters during the present apparition of the comet. There are three main possibilities:

Ha'li rhyming with alley, the obvious one for anyone accustomed to the peculiarities of English spelling.

Hā'li rhyming with bailey, often preferred by those who grew up with the pop group known as Bill Haley and the Comets.

Hö'li rhyming with bawley, favoured by Colin Ronan, one of Halley's biographers, on the grounds that the astronomer's name was sometimes spelt Hawley; but then it was also spelt Hayley or Hally, on occasions.

My assistants telephoned sixteen Halleys living in London to ask them what they called themselves. Three declined to say but every one of the remainder admitted to Ha'li (rhyming with alley), although one respondent mentioned a brother who called himself Hā'li (rhyming with bailey). With such an overwhelming verdict about the present pronunciation of the name there was no point in continuing the survey, especially as one lady was very upset by the question and exclaimed, 'I know *all* about this and I don't wish to comment'.

Halley had his comet and Newton had the universe. The return of the comet in 1758 came to be regarded as a satisfying confirmation of the Newtonian theory on which the prediction was based. But the story has a sting in the tail. It was a dazzling theory that encompassed comets and everything else, and the physicists and astronomers who came after Newton thought it flawless. Every comet that flew, every new planet that turned up, every observation of distant pairs of stars orbiting around each other, nurtured in them a complacency that Newton himself had not shared. 'What I tell you three times is true,' said the Bellman. And where the medieval believers in crystal spheres had been blinded, the Newtonians saw things that never were.

In 1860 Urbain Jean Joseph Le Verrier, the clever and detested director of the Paris Observatory, arranged for the decoration of the Legion of Honour to be bestowed on the village physician who found the planet Vulcan. He was readily persuaded of the doctor's discovery because he had himself deduced the existence of a new planet, to account for discrepancies in the motion of Mercury, the closest of the known planets to the Sun; as one of the predictors of the genuine planet Neptune, Le Verrier fully expected to repeat his success.

During the next two decades German and American astronomers reported seeing the new planet orbiting very near to the Sun, but their reports contradicted one another as to its position. Vulcan was just as non-existent as the moon of Venus, which Giovanni Cassini, Le Verrier's predecessor at the Paris Observatory, had reported in 1672 and many other astronomers duly saw thereafter. The explanation of the peculiarities of Mercury's movements did not transpire until the twentieth century, and it came for Newton's followers as another horrible sight. Stars photographed at the time of a solar eclipse in 1919 shifted their apparent positions in the sky: their rays of light were bent by gravity as they passed the Sun.

Newton had not visualised any such effect, and even if he had done so the deflections of light would, within the framework of his theory of gravity, be substantially less than what was seen. The deflections accorded well with the predictions made in 1915 by Albert Einstein using General Relativity. In short, the eclipse pictures showed that Newton's theory was wrong. The mistake that everyone had overlooked for two hundred years was an implicit assumption that light travelled at an infinite speed.

Another victim of the Einsteinian revolution was Euclid's geometry and, once again, the great author was not himself to blame for the world being bamboozled, in this case for two thousand years. In the neatest piece of doublethink in the history of scholarship Euclid's word 'axiom', meaning just a working assumption, was made to signify the opposite: a self-evident truth. Although it is the job of fundamental research to stop teachers telling so many lies, it is sometimes miraculous that the human imagination keeps functioning at all. The accounts of nature become a little more reliable all the time, but the pay-off in the classroom is slow to materialise.

Sixty-odd years after Newton's theories were superseded, innocent children still learn them by rote as if they were true, and should any youngsters want to become physicists they have to unlearn them. It is only fair to say that the motions of the comet

Halley, cruising in now through Einstein's curved space, are indistinguishable from those it supposedly made under Newton's force of gravity, but that is no more to the point than the fact that navigators still find it convenient to pretend that the Earth is at rest and the stars wheel around the sky. To suggest that Newton's system of the world was only a little bit wrong is like reassuring a corpse that it is only slightly dead. And anyone censorious about errors in the past, and not about his own thinking, deserves whatever may be coming to him in the form of new horrible sights, and I don't mean Halley's Comet.

Astronomy as art: the comet of 1664 depicted in Theatri Cosmicum, *1668.*

3.
THE FABULOUS CLOUD

Giving surnames to pieces of the universe is invidious. Only mutual scorn between the French and British spared us from having to refer to the giant planet beyond Saturn as George rather than Uranus (in honour of King George III, a promising source of funds for research) or to its sibling Neptune as Le Verrier. One nearby star is called Barnard and many inconspicuous stars have names and numbers from their catalogues – Wolf 359 is another neighbour in space.

Craters on the Moon and on Mars bear the tags of scientists. Anyone wanting to see how famous men really stack up in the league tables of history may care to measure off the diameters of craters assigned to them under the auspices of the International Astronomical Union. Visibility depends on other factors, too, and the lunar crater Tycho (84 kilometres wide) is the most conspicuous from the Earth. Nevertheless, the sense of a pecking order is plain: Einstein is 160, Newton 113 and Halley 35 kilometres wide. On the planet Mercury the crater names show that astronomers and space scientists are cultivated folk and they have inscribed, for instance, Mozart (225), Matisse (210) and Mark Twain (140), while the largest crater by far goes to Beethoven (625 kilometres wide) followed by Tolstoy, Raphael, Goethe and Homer, in that order.

This dubious nomenclature is most obtrusive among the comets, because Edmund Halley showed lesser men the road to instant fame. As a result some comets acquire tongue-twisting names like *De Kock-Paraskevopoulos* and *Schwassmann-Wachmann 3*. Although the elders of Cometsville prudently limit the number of surnames per object to three, the appellation can become as long as the comet's tail: for instance *Bakharev-Macfarlane-Krienke*. These cosmic visitors are usually as forgettable as their names, but I regret to say that *Schwassmann-Wachmann 1* never goes away.

There are two ways to pin your name to a comet. For Halley and

A comet discovered in 1893 by William Brooks and photographed by his rival Edward Barnard.

a few others it was a matter of studying the orbits for various appearances of comets and inferring that a single object was responsible for them all. Otherwise it is a simple race to be the first to spot a new one and as a result many comets that might slip by quite unnoticed from the Earth are picked up by enthusiasts with telescopes. Anyone can play this game, if he is willing to leave his bed on moonless nights. Or she: Caroline Herschel, Ludmila Pajdušáková, Liisi Oterma and Eleanor Helin are notable among those who have discovered comets. So we cannot count comet hunting, like warfare and preaching, as simply an aberration of the male sex hormones.

Nor can there be any denying Halley's personal responsibility for this insomniacal form of comet fever. The original fanatic was Charles Messier, a clerk at the Paris Observatory, who watched out for *Halley* on its first predicted return in 1758–9 and went on to become world-famous for discovering new comets. Messier's chief competitors at the end of the eighteenth century were Pierre Méchain in France and Caroline Herschel in England. Comets rattled into the archives like aristocrats on the way to the guillotine. As a 39-year-old porter at the Marseilles Observatory, Jean-Louis Pons spotted his first comet in 1801; by 1813 he had found a dozen and the observatory promoted him from porter to assistant astronomer, but the ungrateful Pons left to become Astronomer Royal at Lucca instead. He discovered thirty comets, and many more than that by the laxer standards of attribution of his time.

Facing: A blaze of glory for the Japanese. The Ikeya-Seki Comet 1965 passed very close to the Sun and developed a long tail. It is one of a family of 'sungrazers' produced by the break-up of an ancestral comet. (Jet Propulsion Laboratory/Table Mountain.)
(Photo in color on inside back cover.)

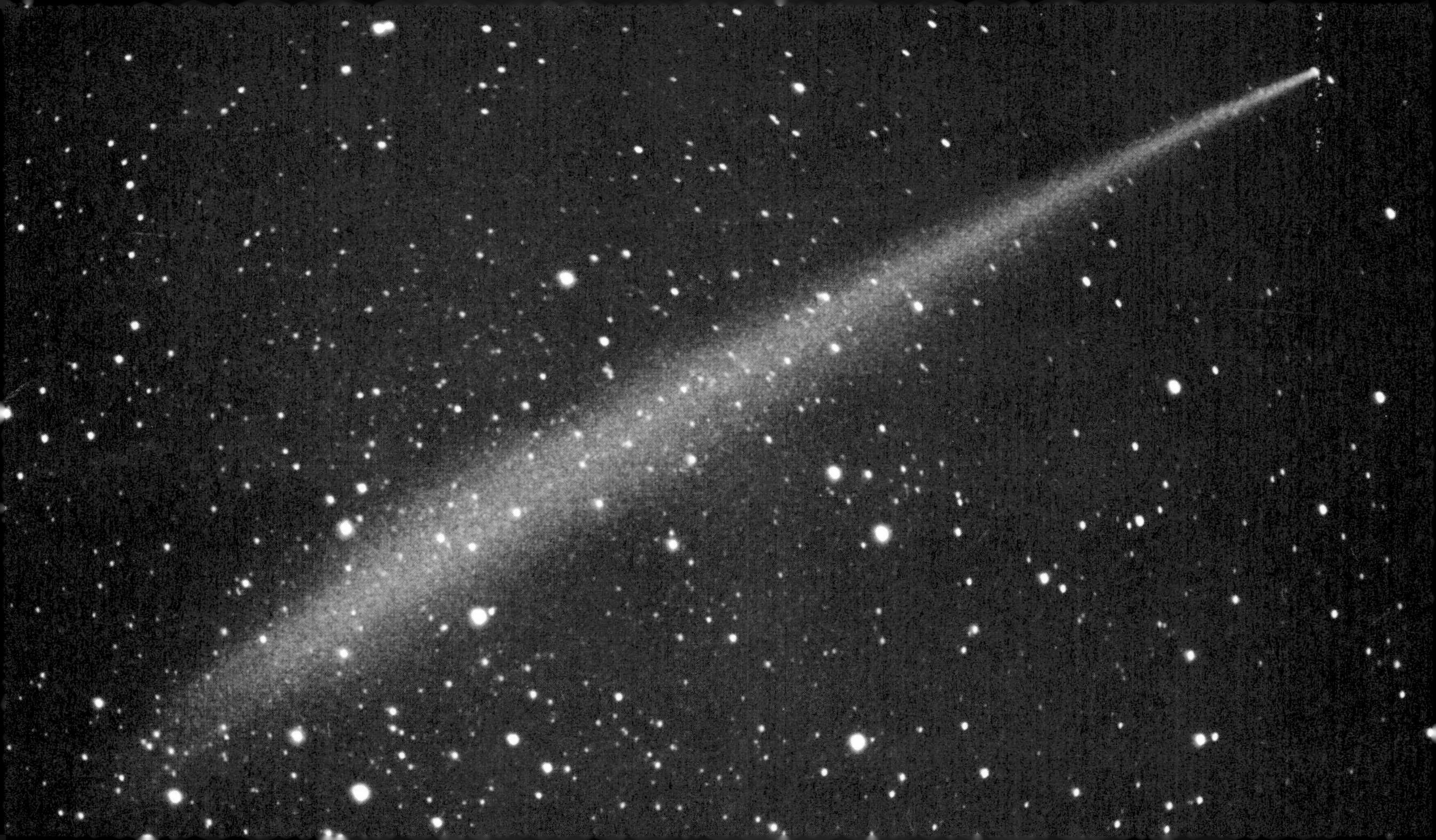

Bennett's Comet 1970 seen over the snow of Switzerland. Reactions to it typified the mixture of se and nonsense that surrounds comets even in the twentieth century. While instruments carried spacecraft were detecting a huge volume of hydrogen gas around the comet, some Egyptians feared was an Israeli secret weapon. (Observatoire de Genève: Photographed at Gornergrat 26 March 19
(This photo shown in color on inside front cover.)

One generous comet increased the hunters' productivity wonderfully. Méchain discovered it in 1786, Herschel discovered it in 1795 and Pons discovered it twice, in 1805 and 1818. A young German astronomer, Johann Encke by name, blew the whistle that ended this free-for-all, by predicting that the comet would reappear in 1822, which it duly did. It was a comet on a very small orbit that brought it back every 3.3 years. All other names were suppressed and it was hailed as comet *Encke*. It is a faint and stubby-tailed thing, yet *Encke* now ranks second only to *Halley* in the eyes of comet lovers and a later chapter will relate how *Encke* became a piebald snowball.

By the nineteenth century most professional astronomers had far more significant matters to study in the sky: new planets, the sunspot cycles, variable and double stars, the chemistry and speeds of stars revealed by spectroscopes, and the layout of the Galaxy as first elucidated by William Herschel, Caroline's brother. But for princes and the general public comets were all that mattered and in 1881 a magnate in the United States offered a cash prize of $200 for any comet discovered from North America.

Enter the comet cowboys, healthy young Americans who saw a short cut to fame and fortune. William Brooks, an immigrant draughtsman, scored particularly well, while Edward Emerson Barnard, a photographer by trade, soon paid off the mortgage on his cottage with 'several goodsized comets' and became a celebrated professional astronomer with a star to his name as well. Barnard also made the first discovery of a comet using a photograph of the sky taken through a telescope, and thereafter other professionals began accidentally picking up comets on plates exposed for other purposes.

For a while it looked as if comet hunting might lose its romance, as twentieth-century astronomy marched ahead to its historic discoveries of the exploding universe and exploding galaxies. To most professionals the triviality of individual comets became obvious. But that left the amateurs and a handful of professional astronomers continuing to enjoy a special relationship with the comet-struck public which, unaware that comets had been downrated, regarded them with an awe that was still at least partly superstitious. Comet hunting remained a high road to prestige, as a Japanese lad demonstrated in 1963.

Kaoru Ikeya was the unskilled son of a failed businessman; the father had turned to drink, leaving the mother to work as a hotel maid. The young Ikeya felt shame, a terrible affliction in Japan, and Peter Lancaster Brown, the Homer of comet hunters, describes his fevered thoughts:

> He needed fame, but how could one so young expect to achieve this? It was impossible. No, not quite. If he could attach his dishonoured name to a new comet, his glory and synonymously his family glory would be rung over all Japan.

Ikeya set to work to make a telescope and he fulfilled his plan as if in a fairy tale: he found his comet and the television crews came hurrying around. He went on to spot others, including the conspicuous *Ikeya-Seki* of 1965.

I do not know Ikeya but I once visited his co-discoverer Tsutomu Seki in the small city of Kochi. This charming teacher of classical guitar told me how he had saved for ten years to buy his special sky-sweeping binoculars; mounted beside them on his roof Seki also had a home-made telescope, its mirror ground for him by a friend, which served for photographing a moving comet from night to night and deducing its orbit. I admired Seki's beautifully embroidered, electrically heated slippers that kept his feet warm at night. But I thought it strange that the names of these comet samurai should resound through the Solar System, while the current discoverers of far greater entities were mute inglorious Newtons: Schmidt who saw the terrible power of the quasars; Penzias and Wilson who found that empty space was still warm after the Big Bang; Hewish and Bell who discovered the pulsating stars; and so on. But a few of them later won Nobel prizes, attaining fame and fortune by a more difficult route.

During the 1970s William Bradfield in Australia broke a Japanese near-monopoly by finding ten comets, while China came back into comet hunting after a lapse of several centuries with a quickfire succession of discoveries. Faithful to maoism, the Chinese professionals scorned to offer the names of individuals to the international authorities; so their objects are named Purple Mountain, or *Tsuchinshan,* after their observatory at Nanking. And while some amateurs, like the village postmaster at Copthorne in England, content themselves with calculating the orbits of known comets and enjoy weeks of harmless numerology with a desk calculator, many others continue to scan the sky, in the hope of writing their names up there. Cash prizes are still awarded in North America and Japan.

Facing: Ikeya's Comet of 1963 brought much-wanted fame to Kaoru Ikeya at the age of nineteen. (Alan McLure, Frazier Mountain.)

Do you wish to enter this cosmic lottery? Point a telescope in the right direction at the right time and you may see a comet

Tsutomu Seki and his rooftop observatory, photographed in 1968, when the Ikeya-Seki Comet of 1965 had made him famous, too.

Facing: William Bradfield, an Australian defence scientist and amateur astronomer, made a remarkable succession of comet discoveries in the 1970s.

before anyone else does. Small telescopes, of the kind that amateurs can make or buy, generally sweep the heavens better than the giant instruments of the major observatories. You will also want a good atlas of the sky to show you what fuzzy objects are *not* comets – but beware of the wandering planets which are sometimes reported as comets, if not as flying saucers. Then you need dark skies, clear weather, a tolerant household, and the patience to spend perhaps hundreds of hours to find one comet. With your photographic memory for what the 'normal' sky looks like when free of comets, you should have to refer to the atlas less and less often. Comets can appear in any part of the sky, but you are most likely to spot one within a couple of hours of sunrise or sunset.

When you have discovered your comet you should send off a telegram without delay. The address, TWX 710-320-6842 ASTROGRAM CAM, obscures the true location of Cometsville, which is where your follow-up letter goes to:

Central Bureau for Astronomical Telegrams
Harvard-Smithsonian Center for Astrophysics
60 Garden Street
Cambridge, Massachusetts 02138

Brian Marsden is the friendly Englishman who runs the bureau on behalf of the International Astronomical Union. As chamberlain

of the comets he will want to know the time and place of your observation, the position, brightness and appearance of the comet, and how quickly it is moving in relation to the stars. If that seems like an expensive telegram, there is a code that you can use.

Before you call the press conference, you should know the chances are about ten to one that you have failed. You may not be seeing the comet *Reader* at all but a well-known nebula or a ghost image in your telescope; otherwise you have been beaten to it and you learn to your chagrin that the comet is already named *Other-Ander-Autre*. The astrogram bureau deals with other sightings besides comets: with apollo objects passing close to the Earth, for instance, and fizzing 'new' stars with which the amateur spotters make significant contributions to stellar astronomy. But the bureau is swamped mainly with reports of new comets, most of them fallacious.

The professional comet specialists know what they are doing, in serving as referees in this never-ending, world-wide contest. Most of the agencies that finance high astronomy would frown on the allocation of salaried astronomers and valuable telescope time to so frivolous a task as looking for comets. An accidental discovery now and again might be forgiven, but nothing as time-consuming as random searches. On the other hand the comet-loving professionals are happy to encourage amateur hunters to give their services at no cost except a handout of glory. Occasionally the searchers will spot a big one coming, which may be worth a glance from the major observatories and the spacecraft. More often the comets become nothing more than statistics which help the professionals to study the social life of comets.

The product of the sleepless nights is a fairly regular count of a dozen or more comets seen from the Earth each year. Of these about half are coming around predictably on known orbits, usually at quite short periods, and half are comets not seen before. Towards the end of 1980 the total score of well-docketed comets in Brian Marsden's catalogue stood at 666, almost all of them harvested since Messier's time. Really bright comets that scare the public out of its wits come only a few times in each century.

All modern theory-spinning about the origins, life-cycles and fate of comets has to begin and end with the statistics of known comets. There are many ways of classifying them, depending on what you are trying to prove. Here I follow a system recently favoured by Armand Delsemme of Toledo, Ohio, although I invert his list, add one to it and put my own tags on the classes.

Kohoutek class: roughly a hundred comets which have come in from distances so immense that even light would take more than six months to cover them: half to one light-year away, or about 35,000 to 60,000 times the distance of the Earth from the Sun. They are distinguishable by their high speeds near the Sun.

West class: about 450 comets are known to arrive on orbits taking anything up to two million years to complete. *West* itself, seen in 1975–6, showed up after an orbit lasting 16,000 years.

Halley class: sixteen comets return to the Sun within less than two hundred years but more than twenty years. None of them strays very far beyond Neptune, the main outpost of the Solar System. *Halley* is the brightest of them.

Pons-Winnecke class: roughly a hundred comets are on small orbits that bring them back at intervals of between five and twenty years. *Pons-Winnecke* itself takes a little more than six years to orbit the Sun and its return in 1983 is its nineteenth recorded visit. But this class suffers high losses from encounters with planets that alter their orbits.

Encke class: as the only comet in its class, *Encke* is in a relatively 'safe' orbit among the inner planets lasting, as I have mentioned, 3.3 years. Its visits to the Sun in 1980 and 1984 are the fifty-second and fifty-third to be observed.

Facing: West's Comet of 1975–6 was a representative of a class of comets that come to the Sun from a very great distance, but not from as far as the Cloud where virgin comets originate. It approached on a 16,000-year orbit, but it was perturbed and the comet has now been virtually expelled from the Solar System. (Lick Observatory.)

The Pons-Winnecke Comet (centre of picture) is at another extreme from West's – a dying comet captured in a tight orbit which whirls it around the Sun every six years. If comets of this class are not expelled by planetary 'football', they will eventually lose their glow entirely and perhaps evolve into dark apollo objects. The streaks in the picture are stars because the telescope tracked the comet moving in relation to the stars, for a quarter of an hour. (G. Van Biesbroek, Yerkes Observatory, 1927.)

Nearly all in the first category, the *Kohouteks*, are judged to be 'new' or virgin comets approaching the Sun for the very first time. I put 'new' in quotation marks because comets are generally supposed to have existed for billions of years before we see them. The last in the list, the *Pons-Winneckes* and *Encke*, are 'old' decaying comets. A plausible line of reasoning links all these kinds of comets, and accounts for the statistics, in a single scheme. But the reader would be left with far too unmuddied a view of the origin of comets if I did not first mention some of the competing theories.

Curiously forged from the ancient idea that comets were exhalations of the Earth's own atmosphere, there has been an unbroken chain of belief among some astronomers that comets are flung out of planets. For example, Johann Hevelius in 1668 reasoned that planets were made of perfect aether, but had to spin to be everlasting, otherwise the heat of the Sun would damage them. Less than perfect exhalations from the planets, mainly Jupiter and Saturn, were hurled like sling-shot across the Solar System.

By the nineteenth century, planets themselves were regarded as less than perfect, and volcanoes on Jupiter and Saturn were blamed for throwing out comets. When critics complained about the enormous speeds that nascent comets would require to escape from the gravity of such massive parents, the moons of Jupiter and Saturn were offered as alternative sources. The chief advocate late in the twentieth century of a planetary origin for comets is a Soviet astronomer S. K. Vsekhsvyatsky. In 1979 a visiting American spacecraft sent back pictures of a volcanic eruption on Io, a moon of Jupiter, but there was no sign of any comet coming out.

Another version of planetary origins was detailed in 1978 by Thomas Van Flandern of Washington, who said that the comets came from a fairly large planet that formerly orbited between Mars and Jupiter and broke up only five million years ago. Most of its substance left the Solar System entirely but chunks of its core remain in place, in the main belt of asteroids between Mars and Jupiter. The present-day comets are pieces that were flung out a long way and are now falling back again. In this story we are troubled by comets only because we have chosen the wrong time to be alive, but not quite as bad a moment as five million years ago, when the Earth was supposedly spattered with rocks and water from the exploding planet and comets abounded for thousands of years. If there were any good reason for crediting Van Flandern's theory, we should be congratulating ourselves on a narrow escape.

Interstellar space is the source of comets, in another sort of theory which I call the heavenly hoover. In France in the early nineteenth century, while Joseph-Louis de Lagrange was championing planetary origins, his rival Pierre-Simon de Laplace was advocating the capture of comets from clouds existing between the stars. In England in the 1950s Raymond Lyttleton saw the Sun focusing the interstellar dust with its gravity. In the latest version, from Scotland in 1979, William Napier and Victor Clube picture the Sun running, every thirty to fifty million years, through one of the spiral arms of our Galaxy, the Milky Way. There it encounters dust clouds which stock the Solar System with fresh supplies of ready-made comets and other small but less luminous pieces of cosmic matter. The last episode of cloud-winnowing ended no more than ten million years ago, so they say, and the present supply of comets is rapidly declining.

The statistical data on comets, the fruit of all those night watches, come into the argument here. If the Solar System episodically recruits 'new' comets you might expect the directions from which they arrive to show some bias related to the direction of the Sun's motion through space. Napier and Clube defy the general belief that 'new' comets arrive randomly from all directions in the sky and say that there is a significant bias; they invoke disturbances by other stars to explain why the bias is not more pronounced. Any strength in this idea lies in the reasonable supposition that *some* 'new' material must come into the Solar System from interstellar space, if not from dust clouds then from families of comets belonging to other stars. The heavenly hoover suffers, though, from one overwhelming weakness: it is unfashionable.

If the 'new' comets seem to be coming from starting points half a light-year out, that is where they have been in residence ever since the birth of the Solar System. Such is the leading idea about the origin of comets, that is to say the theory not disfavoured in Cometsville. All the other categories of comets are, in this scheme, pilgrims from the coolest provinces of the Solar System who have lost their return tickets and are trapped in shorter orbits around the Sun, waiting either to perish in the heat or to be evicted into interstellar space and exiled from the Sun for ever. But the theory requires the existence of a large population of unseen comets to sustain the pilgrimage, and thus it endorses the opinion of Johann Kepler that 'there are as many comets in the sky as fishes in the sea'.

The devotion of comets to the gravitational faith that unites the Solar System appears in this: their first journeys from the outer darkness to the altar of the Sun take a very long time indeed, for the sake of a fleeting visit. When the priests of this faith, the celestial mechanicians on the Earth, interrogate 'new' comets now arriving in the Space Age they admit to travelling for several million years. They took the plunge before the recent series of ice ages began on the Earth, when our ancestors were romping on the African savanna, a couple of species before *Homo sapiens*. And for this reason alone, the long travel time, the comets begin to proliferate outrageously.

One or two 'new' comets appear each year in the inner Solar System and there may be several more that escape detection. If that supply is not about to cease, you must suppose that millions of comets are already on their way, coming in from the cold. The object that is going to put on a show a million and one years from now is well into its pilgrimage and this implies the existence of a swarm of comets converging on the Sun from all directions. But they begin their journeys extremely slowly, at a few metres per second only, no faster than a camel, because the Sun's gravity is very weak half a light-year out and the comets have yet to gather speed in their long drop. Most of the expected comets are still bunched near their starting points, in a shell-like cloud between half and one light-year out. And there, on the supposition that only a very small minority of comets ever visit the Sun at all, the reasoning multiplies their numbers yet again.

Ernst Öpik is an Estonian astronomer and musician who has recently been running the Armagh Observatory in Northern Ireland. For most of his long life he has adopted the role of cosmic garbage-sorter, concerning himself with the stray material of the Solar System. In 1932 he calculated that an invisible cloud of comets and meteors, surrounding the Sun at enormous distances, could survive throughout the long lifetime of the Solar System. In 1950 the doyen of Dutch astronomers, Jan Oort of Leiden, who is better known for classic work on the nature of galaxies, reworked Öpik's idea. He emphasised a different aspect of it, namely that passing stars would cause a few of the objects to fall out of the cloud and into the heart of the Solar System, to become observable as 'new' comets.

Thus was the fabulous Öpik–Oort Cloud conceived, as the source of the comets. I abridge the name to the Öoo Cloud and defend this coinage on grounds of sight and sound. It looks like an untidy collection of roughly round objects of various sizes, and it is pronounced 'Er, oh!' – just what a neophyte comet lover is liable to utter when he is first told that there are many billions of the things out there.

The Öoo Cloud is fabulous in both senses of the word: amazing on the one hand, totally hypothetical on the other. As critics are quick to point out, it is by definition as invisible as a cloud of gnats a million miles away and there is no possibility of observing it directly – until perhaps the first starship leaves the Solar System and passes through the Cloud on the way out. But if scientists were obliged to stick to what they can see, they would have nothing to say about the interior of the Sun or family life among the dinosaurs, and we should have no electronic gadgets at all, because electrons are quite invisible. So I shall describe the Öoo Cloud on the supposition that it is real. The numerous comets are very slowly orbiting the Sun, coming no closer than the outermost planets and spending by far the longest phases of their existence at much greater distances, in the Cloud. Then you bring in the passing stars.

Edmond Halley himself first discovered 250 years ago that the stars which look so steady to the casual observer are, in reality, whizzing about at high speeds in relation to the Sun. For those of a nervous disposition who suspect, quite rightly, that a collision with a star would be even more disagreeable than a collision with a comet, let me say that the sky is a big place: during a ten-billion-year lifespan of the Solar System the closest that any star is likely to come is about five hundred times the distance from the Earth to the Sun. But the Öoo Cloud is a hundred times farther off than that, and as often as once in a million years a star will charge through a segment of it like a wild bull through a garden party, scattering the guests.

Perturbed by the star's gravity, many millions of comets take flight, quitting the Solar System for ever. Others change their orbits drastically and yet remain in the Cloud. And ten million comets, for the sake of argument, are stopped in their tracks like bull-watchers paralysed with fear. As it was only their slow orbital motion that kept them at their distance from the Sun, these arrested comets can then do nothing but begin the long, long fall. But they will not be bunched: a number of star-encounters will occur during the time of several million years required to complete the fall, and any small residual speed (up, down or sideways) that a comet possesses at the start makes a huge difference to the time and direction of its arrival at the Sun.

The action of a star hurtling through the Öoo Cloud can be regarded as a special case of gravitational football, in which comets are booted about the sky. Those that fall into the heart of the Solar System become playthings of the planets, which switch them from orbit to orbit. To understand the rules of this game, consider first the simple picture of a comet plunging down to the Sun, swinging around behind it, and then climbing away, more or less on the same side of the Sun as that from which it approached. Half close your eyes and you might say it had simply bounced off the Sun's gravity. Next in your imagination, gently bounce a football off the front of a moving car and note that it will shoot forward at twice the speed of the car; bounce it off the back of the car and the football will come almost to rest. Physicists who speculate about interstellar travel envisage the spaceship skipper of the future gaining enormous speed at no cost or strain by bouncing himself off a fast-moving star – or, more precisely, swinging his ship around it, like a comet around the Sun. He could later reverse the procedure. The comets that topple out of the Cloud have been slightly slowed by remote effects of passing stars.

The planets, too, are swift gravity-spheres, from the viewpoint of a passing comet or a spacecraft, and some of the neatest operations with deep space probes have already used a planet's motion to redirect the spacecraft's motion, passing it into a new and faster orbit. Similarly a close encounter with a planet's gravity can greatly change a comet's speed and direction, altering its orbit around the Sun. Often it will cross a planet's path roughly at right angles: if the comet passes in front of the planet it will lose some of its energy of motion, if it passes behind it will gain energy and that is really better. In many cases the planets will kick the comets right out of the Solar System, which is the cleanest way of getting rid of them.

As the largest planet and chief defender, Jupiter does most of this work, but any of them, the Earth included, will lend a boot now and again. About half of the 'new' *Kohoutek*-class comets are expelled at their very first foray to the Sun: that is to say, they gain energy and, when they return to the Öoo Cloud, they will just keep on going, out into interstellar space. The other 'new' comets (*Kohoutek* was one) lose energy and have their orbits reduced, either to a moderate degree, making *West*-class comets, or drastically, as in the case of *Halley* and its kin.

The comets now spend more of their time in the vicinity of the planets, vulnerable to further encounters, which may still expel them, or at least put them into large orbits out of sight. *West*, for instance, gained energy on its most recent visit and is due to quit the Solar System. In other passes they may lose more energy of motion and join the *Pons-Winnecke* class on small orbits. This football is not hearsay: astronomers can sometimes infer particular events, as in 1886 when the comet *Brooks 2* overtook Jupiter and closed to within about 100,000 kilometres of the planet's surface. The comet swung around in front of Jupiter and 'bounced' back. That greatly reduced its speed in relation to the Sun and the period of its orbit was cut from twenty-nine to seven years. Jupiter's moons, on the other hand, were quite unperturbed by the comet rushing among them.

The *Pons-Winnecke* comets, visible so often from the Earth, are also at frequent risk from encounters with Jupiter, as well as from wastage and disruption during their repeated visits to the Sun. What puts the comet *Encke* in a class of its own is that it tucks the far end of its orbit safely inside the orbit of Jupiter. It seems to have done this using a self-braking system to be revealed later.

Another lesson from the statistics of comets is that, although the 'new' ones come in from all directions in the sky and half of them go the 'wrong' way around the Sun, contrary to the sense in which all the planets revolve, the other classes are confined progressively closer to the sheet of space in which all the planets lie. And by the final *Pons-Winnecke* and *Encke* stages all the comets are circulating in the 'right' sense around the Sun. This is the best evidence for an evolution of comets through the various classes. It can be understood by the rules of cosmic football, because the effect of encounters is to make the orbits of comets that survive in the inner Solar System more and more like the orbits of the planets.

To keep these clumsy processes going and account for the few hundred visible comets, the theorists are obliged to stock the Öoo Cloud with about 100,000 million comets. Even in such a multitude, the total mass is less than the Earth's, which shows what slight things comets are. Astrophysicists also suppose that, at the origin of the Solar System, there were many more than that. In this scheme the comets are a long-lasting by-product of the birth of the Sun and its planets.

Nobody saw the events some 4600 million years ago that constructed the Solar System but any astronomer worth his salt will spin you a good yarn about it. Evidence that he may even be right comes from telescopes that show new stars and planets forming out of the loose clouds of gas and dust in the Galaxy. The commonest materials in the clouds are the gases hydrogen and helium, which are also the principal constituents of the Sun and its giant planets, but there is a rich admixture of grains of ice and dust, containing every other chemical element. The usual account has such a cloud collapsing faster and faster under gravity to make a swirling disk of gas, ice and dust, with the proto-Sun at its centre. In an important new variant of the story, it seems that a pre-existing star exploded immediately before this happened, an event that has some bearing on the mechanism of cloud collapse and the details of cometary constitution.

The energy released by swiftly infalling material at the core of the Sun heated it sufficiently to start the thermonuclear reactions that were to keep it burning for billions of years. Newborn stars throw off much of their substance in a violent wind, so, in the surroundings of the young Sun, what happened was the result of the combined effects of gravity, heat and the solar hurricane, acting over different time-scales on the agglomerations of gas, ice and dust in the disk. In some versions magnetism plays an important part, too, but I must sidestep many detailed issues that keep astronomers who play God arguing into the small hours. The general picture is of swarms of stony and icy lumps growing, colliding, breaking up and gradually gathering together.

Many of the supposed 'planetisimals' bear an uncanny resemblance to comets, and astronomers who say that the comets built the planets are most convincing about the icy planets, Uranus and Neptune. The biggest of the newborn planets, Jupiter and Saturn, gathered a great deal of the available hydrogen. Gassy and icy material could not survive close to the Sun, so the inner planets including the Earth were made of stony stuff. The comets of the Öoo Cloud may have originated mainly in the regions where Uranus and Neptune were constructed and, if so, they represent waste material from that operation.

As each planet grew it became a bigger and more powerful attractant for further raw materials, but at the same time its gravity made it a formidable footballer, and comets that experienced near misses were liable to be booted out of the Solar System. A minority of them were punted away just far enough to set them up in the Öoo Cloud. Nature could afford to be extremely wasteful in this process: comets are such slight objects that the mass of the entire original population of the Öoo Cloud was probably a good deal less than either Uranus or Neptune, and in any case far more material was lost from the disk around the Sun than went into all of the planets, moons and comets.

But the football works both ways, and vast numbers of comets were deflected into the inner Solar System, where they ran riot. Many of them collided with the planets. From planetisimals that were probably mainly comets or their remnants, the newborn Earth and its sister planets were subjected to a torrential bombardment compared with which the Vietnam War was a gentle shower.

The Age of Comets has a grand sound and may quicken a cometophile's pulse-rate, but it should really be thought of as a mopping-up phase. After the planets were essentially complete the work of purifying the Solar System took hundreds of millions of years. The solar hurricane blew, sweeping vapours before it, and the pressure of sunlight drove the fine-grained debris out of the Solar System. Nevertheless the sky was full of hairy tails and, as chunks of ice and stone travelling at many kilometres a second came raining down on them, the harder planets and moons were spattered with large craters. On the Earth, geological activity healed the comet-pox scars, but those bodies whose faces froze 4000 million years ago still show them. The Moon, Mercury and Callisto (a moon of Jupiter) are prime examples of the impact cratering in the Age of Comets and calling the craters Einstein or Beethoven does nothing to cure the disfigurement.

Because so many comets were stored up in the fabulous Cloud, the work of keeping the sky clean is never ending. Sweeping up comets is unhealthy for dinosaurs and other living things, and that will be an important theme of later chapters. But first we should examine more closely the nature of the comets themselves. If you could see the comets in the Öoo Cloud they would probably look like bare, grubby snowballs, drifting at walking pace and quite devoid of fiery heads and tails. Earthbound observers of comets arrived at this simple mental picture by the usual circuitous route.

Facing: A wound in the Earth caused by a cosmic impact is clearly visible in a satellite picture (top) of the crater at Manicouagan, Canada, 70 kilometres wide and 210 million years old (ERTS picture by courtesy of R. Grieve). In the lower photograph, from the Danish coast near Copenhagen, 'the coffin of the dinosaurs' is the dark band between light-coloured rocks below (Cretaceous) and above (Tertiary). As explained in Chapter 7, this layer has a peculiar atomic composition and is thought to consist of dust thrown up by a comet or apollo object hitting the Earth. (Photograph by courtesy of J. Smit, University of Amsterdam.)

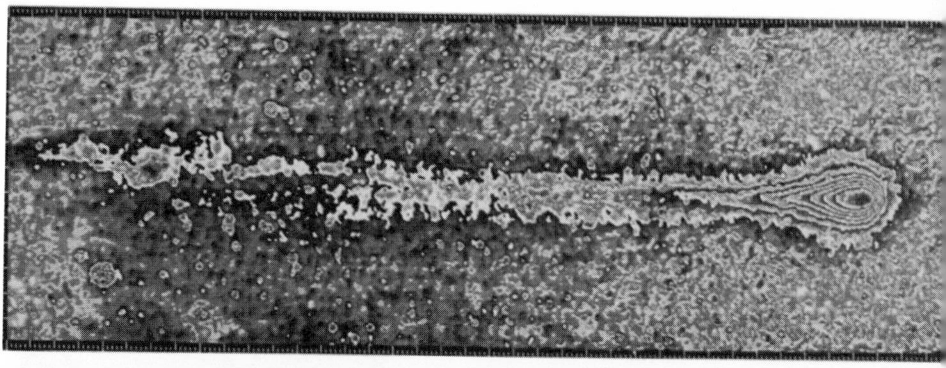

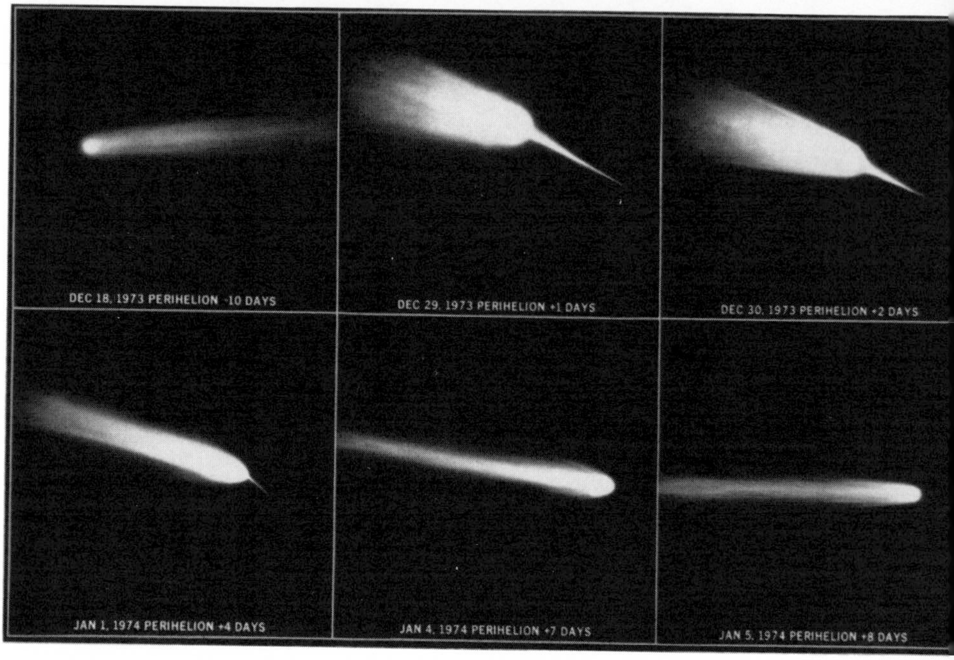

Kohoutek's Comet 1973 received some of the modern methods of scrutiny that Halley's Comet 1986 u undergo. The uppermost image is a Christmas Day photograph from the Skylab space station that u processed by computer to represent the luminosity of different parts of the comet by contours and fa colours. (J. Lorre, JPL/IPL.) The lower illustrations are an artist's representation of visual obs vations of Kohoutek's Comet made by astronauts in Skylab, just before and after the comet rounded t Sun (at 'perihelion') on 28 December 1973. During this period the comet was not seen from the groun Note the anti-tail in the second picture. (NASA.)

(These photos shown in color on inside front cover.)

4.
HEADS AND TAILS

During the comet *Halley*'s last visit to the Sun, in 1910, reputable observers said that it broke up. At the time this was widely believed to be true but, after the comet departed, calmer opinions prevailed and the reports were seen as the products of imagination working on certain jets and streamers visible in the comet's head. Like the planet Vulcan, it was an example of people seeing what they expected to see: there would have been no such suggestion a hundred years earlier.

An amiable mathematician told me how he once thought Saturday was Sunday. All the evidence of Saturday bustle around him failed to shake his conviction and he rebuked an astonished newspaper-seller for fobbing people off with yesterday's papers. Only in the evening, when he went to deliver a society lecture and found no one there, did the truth begin to dawn. Again, the annals of war are full of instances of troops, ships and aircraft being misidentified and efficiently destroyed by 'friendly' forces. Christopher Columbus's belief was that he had sailed west to Asia and we all connive in it to this day by calling the American natives 'Indians'. And there is no reason to doubt the sincerity of those who spot flying saucers.

A rich vein of experimental psychology was opened in the late 1940s when Jerome Bruner of Harvard began researching into influences on human perception. In a series of classic experiments with Cecile Goodman and Leo Postman, he showed (1) that valuable coins look larger than they really are, (2) that emotionally loaded words like 'crime' and 'bitch' are harder to read than bland words, and (3) that when people are shown a false playing card, say a red six of clubs, they are likely to report it as a six of hearts. In this playing-card experiment a person may hesitate uneasily for quite a while before he sees the deception and typically exclaims, 'My God!' The historian of science Thomas Kuhn has drawn the parallel between this experiment and the reluctant shifts of ideas in science.

Expect strange sights in the sky and you are liable to see them. A medical student, Ambroise Paré, wins pride of place in the chronicles of wishful seeing for his observations of the comet of 1528. He was an able young man who went on to be surgeon to four kings of France and invented, among other practical novelties, ligatures for stopping arterial bleeding. Yet in all honesty he described the comet as being bloody in colour, which is plausible, but also stocked with weapons and arrayed with hideous human heads, which seems unlikely.

A malodorous comet demonstrates even better how ideas influence perceptions. Medieval scholars in Europe, who deferred to ancient wisdom and considered that comets inhabited the upper atmosphere of the Earth, also suspected them of being Devil's work. So it is not altogether surprising to learn that certain medieval monks swore they smelt a comet, or that the odour was of suitably devilish sulphurous gas. This report of halleytosis won approval in a book already mentioned, *Cometomantia* of 1684, which used it as evidence against those over-zealous astronomers who wanted to displace all comets into deep space, far beyond the orbit of the Moon.

In the matter of the break-up of comets, the first psychological effect to manifest itself was blindness of the kind that prevented the Europeans seeing 'new' stars: it was followed by the eager supersight which affords a grand view of invisible objects and processes. An historian of ancient Greece, Ephorus by name, reported that in 371 BC a comet broke into pieces; and he was reviled by Nero's tutor, Lucius Seneca, as an irresponsible gossip. Even though Johann Kepler had also said that comets could break up, by the nineteenth century the possibility was either forgotten or regarded as a grotesque fiction. Multiple tails of comets were seen often enough but multiple heads were unthinkable.

When, in the winter of 1845–6, a comet called *Biela* became oddly pear-shaped and then divided into two distinct comets, one of the astronomers who observed them, James Challis of Cambridge, averted his gaze. A week later he took another peep and *Biela* was still flaunting its rude duality. He had never heard of such a thing and for several more days the cautious Challis hesitated before he announced it to his astronomical colleagues. Meanwhile American astronomers in Washington DC and New Haven, equally surprised but possibly more confident in their own sobriety, had already staked their claim to the discovery. Challis excused his slowness in reporting the event by saying that

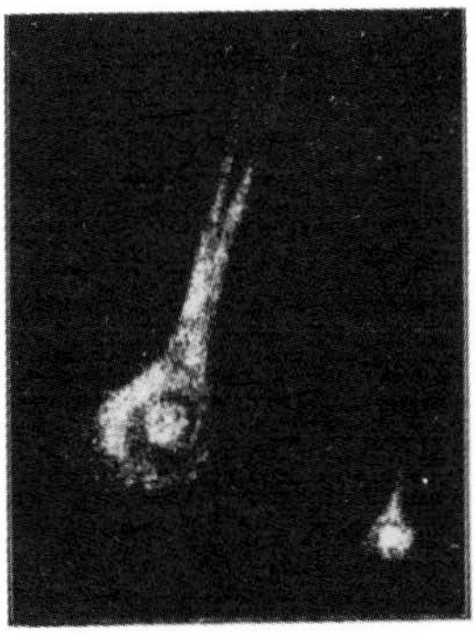

A sketch of Biela's Comet when it had split in two in 1846, to the amazement of astronomers. (After drawings made by a German astronomer at the Pulkova Observatory in Russia.)

he was busy looking for the new planet beyond Uranus. When later in the same year he was needlessly beaten to the discovery of that planet (Neptune) by German astronomers, Challis explained that he had been preoccupied with his work on comets.

After *Biela*'s fission similar sights became commonplace, not because nature embarked on a frenzy of comet smashing but because the astronomers' visual neurones were at last prepared to register such events. In fact, splitting comets became the popular thing to see, whether or not they really happened. For example, the great comet *Donati* (1858) was reported as coming to pieces but it almost certainly did not do so. And inevitably the most famous comet, *Halley*, was supposed to conform with the latest fashion in 1910, and to fall apart. A considered, and one might say considerate, view of those reports is that *Halley* was on the verge of disruption in 1910, but survived intact.

On the other hand, several other comets certainly have split since *Biela*, some of them being well documented in photographs. Zdenek Sekanina, a Czechoslovak-born astronomer in Cometsville, has put the stamp of certified break-up on *Sawerthal* (1888), *Campbell* (1914), *Whipple-Fedtke-Tevzadze* (1943), *Honda* (1955) and *Tago-Sato-Kosaka* (1969). The ones with cumbersome names deserve to split up, but that cannot be said for *West* (1976). This was the great comet that the public did not see, because news editors, feeling foolish after they promoted *Kohoutek* of 1973 as the spectacle of the century, virtually banned any mention of comets from their papers and television screens. The head of *West* divided itself into four pieces and in the process threw out streamers of dust every two or three days, which gave the comet a fanned, peacock-like tail. Paolo Toscanelli described *Halley* in those terms, in 1456, but any inference would be rash.

In the twelfth century, by Brian Marsden's retrospective calculations, a comet passed extremely close to the Sun and broke up into at least two pieces. (It may have been a conspicuous comet of AD 1106.) One of the pieces came back in 1882 as the 'Great September Comet' and again grazed the Sun; another, on an almost indistinguishable orbit, appeared in 1965 as the comet *Ikeya-Seki*. There is in fact a group of 'sungrazers', sharing very similar orbits, and they have been spawned over many centuries by successive break-ups, from a single parent comet. The 'Great March Comet' of 1843 was one of them, and so were at least seven since then. *Pereyra* of 1963 passed within a mere 60,000 kilometres of the bright solar surface, while the 'Great Southern Comet' of 1887 may have collided with the Sun.

The visual system is adept at filling in details in a cartoon, seeing the Man in the Moon and generally perceiving unreal patterns. The ink-blot tests with which psychologists used to plague people are a case in point; another is the interpretation of comet haloes and streamers as bleeding heads or splitting heads. The most celebrated discovery in the history of astronomy was the Canals of Mars. In 1877 Giovanni Schiaparelli of Milan reported that he could see channels crossing the face of that planet and by the end of the century Percival Lowell, founder of a famous observatory at Flagstaff, Arizona, had mapped an intricate network of dozens of canals.

The martians' planet-wide irrigation system was, for Lowell, the dismaying sign of an advanced civilisation fighting the growing aridity of the planet. The American comet hunter, Edward Barnard, wittily disposed of the fiction: he operated what was, at the time, the most advanced telescope in the world, and he said that it was too powerful to show the Canals of Mars. Yet (there is always a 'yet' in science) the ghosts of Schiaparelli and Lowell can have one last snigger. In 1971 a spacecraft sent back pictures of Mars showing huge valleys, including a natural rift eighty kilometres wide and five thousand kilometres long – one of the features mapped by the 'canal' enthusiasts. Of the martian civilisation there was of course no sign, but the channels were not all imaginary. And there were also immense volcanoes on Mars: the jesting Barnard had seen them himself, but did not dare mention them for fear of ridicule.

Some illusions are deep-seated in the machinery of our brains, which compel us, for instance, to interpret the sky as a flattened dome, so that the Moon and the comets always look much bigger on the horizon than when they are overhead. Telescopes, too, can deceive and many an eager amateur mistakes a stray reflection in his instrument for a new-found comet. An observatory in tsarist Russia had a telescope that was well adapted to the discovery of companions of the stars, because a flaw in the glass made it see everything double. But when misperceptions and misconceptions ferment together the results can be more vivid than that.

The discoveries of Isaac Newton and Edmond Halley did not at once dispose of devilish connotations of comets. On the contrary, the realisation that comets move in great ellipses around the Sun helped to modernise the notions of Hell. What better punishment for sinners than to be condemned to ride on a comet for ever and experience a fearful alternation of freezing in the outer darkness and roasting when the comet brushed past the Sun? The cheerful idea that the comet of 44 BC was Caesar on his way to Heaven was thus revised to stock comets with the souls of the damned, not on their way but there already, in a brand-new, eighteenth-century, whirly version of Hell.

Astronomers had more sense than that. They knew that comets were free-ranging lumps of light that were necessary to sustain all living folk, like those who made their home on the Sun. Nowadays the bad habit of attending to only those past ideas that pointed more or less directly towards modern scientific theories is being corrected by historians of science, and from these professional investigators of theories long since dead we learn what creative errors surrounded the attempts to explain the illuminations of the sky, two to three centuries ago. By then the misunderstandings about light were no longer quite as bizarre as the ancient proposition that eyes acted like lasers and looked around by sending out beams, as in the 'darting glances' favoured by some novelists. But there was a persistent confusion between light and its sources. Most of us nowadays can distinguish a lamp from the dust of photons that it throws in our eyes. Our predecessors, though, were infected with the old concept of 'fire' as an element in its own right, and in any case, not knowing about plasmas heated by nuclear reactions, they found the light of the Sun and the stars mysterious.

Even Newton, who founded the modern science of light in his spare time and certainly knew the difference between rays and sources, was at a loss to say how the Sun could burn undiminished for thousands of years (billions of years, as we now know). The comets came like cosmic cavalry to the rescue: Newton said that

God arranged for the Sun to be reinforced by comets that fell into it as a result of fatal changes in their orbits. Here too, he thought, was an explanation for the new star seen by Tycho Brahe in 1572. Astrophysicists today would say it was a natural H-bomb going off, in the destructive explosion of a giant star; for Newton it was a star that suddenly acquired new splendour thanks to a goodly stoking of comet-fuel.

Newton was no Newtonian, in the austere nineteenth-century sense of one content to contemplate the celestial clockwork. The taming of imagination by algebra dates not from the *Principia* of 1687 but from Karl Friedrich Gauss's *Theoria Motus Corporum* of 1809. In the intervening century astronomers, clergymen and laymen continued to regard comets as manifestations of divine purpose. Taking their cue from Newton, they wondered about the light of comets and about their life-giving properties, whether in watering the Earth or conveying some vital principle that might breathe life into all the orbs of the sky. Such ideas foreshadowed a twentieth-century proposition that the Earth might have been seeded with the first bacteria by a comet's tail according to speculations reserved for a later chapter.

The cheekiest of all ideas about the cosmos was William Herschel's belief that the nearest star was peopled: '. . . we need not hesitate to admit that the Sun is richly stored with inhabitants.' Herschel flourished a hundred years after Newton and his name is one of the most celebrated in the history of astronomy. He was a Hanoverian musician living at Bath in England, and was catapulted to fame when his telescope revealed the large planet Uranus, the first major piece of the Solar System to be noticed since prehistoric times. Many astronomers had already seen the planet but mistook it for a star. In the post-Halley era Herschel naturally announced his discovery as a new comet, and it fell to others to infer that it was a planet lying far beyond Saturn and a hundred times larger than the Earth.

Concerning comet light and its contribution to the habitability of the Sun, Herschel pictured alien creatures going about their business on the solid surface of the Sun which underlay an atmosphere of light that was replenished, as Newton had said, by an influx of comets. Cosmic light was thus a special kind of life-giving stuff that nature could manipulate, in this case into a shell around the Sun that illuminated the solar surface as well as the Solar System. Even in Herschel's day, to say that the Sun was cool could earn you a certificate of insanity, but he supposed that the solar rays generated heat only when absorbed in a suitable medium, such as the Earth's atmosphere. And he offered visible evidence in support of the proposition: what were the familiar dark sunspots but chinks in the light offering glimpses of a cooler surface below? Present-day reckonings of the internal temperature of the Sun guarantee that any inhabitants of that world must be, to say the least, jumping like fleas on a hotplate, and sunspots are known to be magnetic storms.

Yet the adherents of these ideas about light are entitled to more than a posthumous chuckle, because their line of reasoning led them to conclusions that the latest astrophysics regards as sensationally correct. The marshalling of the supposed light-stuff around the heavy Sun indicated to eighteenth-century astronomers that light was subject to the action of gravity. The luminiferous comets, moving in obedience to Newton's law, provided equally graphic proof. But the ability of gravity to bend light and slow it down was to be the basis of General Relativity. Albert Einstein arrived at it armed with less devious notions about the nature of light, and the anticipation might be dismissed as a chance intersection of two very different lines of thought, had it not also led to General Relativity's most breathtaking deduction. The eighteenth-century theory of light predicted black holes.

The man to say so first was a clergyman at Thornhill near Leeds: John Michell. At Cambridge he had initiated the modern study of magnetism and of earthquakes; he also identified the double stars and gave Herschel a correspondence course in telescope-making. Michell set down the idea of an invisible star in a paper that was published in 1784 and resurrected in 1979 by Simon Schaffer, an historian at Cambridge. The original black hole arose in Michell's mind from the thought that a large and massive star would, by its powerful gravity, slow down the light that it emitted. In the extreme case a star could choke off its light entirely, and Michell calculated that a star with five hundred times the Sun's diameter, and the same density, would be invisible. Modern black-hole theory gives the same answer. That is just a happy coincidence, as the mathematical viewpoint of General Relativity is entirely different from Michell's. In 1796 the French theorist Pierre-Simon de Laplace arrived at the same thought and similar calculations. He has often been credited with originating the idea of black holes, but it turns out that the Rector of Thornhill was there before the great Laplace.

Astronomy is thus a four-legged animal standing on sound and false ideas at the front and sound and false observations at the rear. Amazingly the beast can limp forward, sometimes even gallop, from one discovery to the next. And lurching on its way to valuable information about the nature and history of the universe it has in passing exposed the character of those ephemera, the tails of comets.

Hold out a green pea and you can eclipse the Moon with it – but don't try it with the Sun or you will damage your eyes. The astronomers of ancient China measured their comet tails in feet and indeed a two-metre ruler, at arm's length, would not span some long-drawn-out comets that stretched across more than half the sky. Comet tails are very variable in length and appearance, from comet to comet and within the period of observation of one comet, so it is no wonder that the astrological secret service in China could not make up their minds whether the tails of comets were like handles or brooms or candleflames or sailing boats. After eyeing these appendages for a few centuries they realised that the tails always tended to lie away from the Sun: that means a comet travels roughly head first as it approaches the Sun and roughly tail first as it moves away. When medieval Europeans overcame their fright and looked carefully at comets, they came to the same conclusion – and then overstated it by saying that the tails pointed exactly away from the Sun, which was wrong.

Imaginations worked overtime to account for comet tails. Some in China visualised comets as side-lit objects whose appearance, to observers on the Earth, changed much as the Moon's does, according to how the Sun's rays fell on it. At the dawn of modern astronomy in Europe, the tails of comets were likened to shafts of sunlight streaming between clouds. One idea on offer in the seventeenth century was that comets were a matter of levity: the tail was said to be a stream of levitating material subject to a supposed force of anti-gravity that repelled it from the Sun. Or the comet allegedly stirred up the aether in cosmic space like a ship ploughing through the sea. Newton pointed out contemptuously that this 'crumbling of the heavens' did not accord with the changing directions of the tail.

From his observations of the 1680 comet, Newton noted that the tail increased as the comet rounded the Sun and the great heat, he supposed, vaporised some of the comet's material. That is a crucial point in understanding comets: the Sun's heat provokes an otherwise unnoticeable object to give off the effusions that leak

West's Comet breaking up in March 1976, in a succession of photographs taken over sixteen days, with an exposure that showed only the brightest parts of the comet's head. Some of the fragments were small but brilliant, presumably because they exposed new surfaces of volatile material to the Sun's rays. (New Mexico State University Observatory.)

An image of the peacock-like tail of West's Comet, processed by computer to sharpen the 'multiple streamers' associated with the break-up of the comet's head. (Big Bear Observatory and Los Alamos IPG.)

across the sky and make it visible. Newton also believed that, much as comets' heads stoked the Sun, so comets' tails were the source of heaven-sent moisture and vital spirit, necessary for the sustenance of planetary life. He went on to muff his own explanation of the tail of a comet, which he likened to smoke rising against gravity in the hot, rarefied air of a chimney. He thought that particles heated by sunlight warmed the aether around them, which then lifted the particles away from the head of the comet.

The first person to interpret the comet's tail correctly, as judged by the examining boards of present-day astronomy, was Newton's predecessor Kepler, who pictured the pressure of the Sun's rays sweeping material away from the atmosphere of the comet's head. That matches modern opinion, according to which the tail consists, in part, of a smoke of dust particles of various sizes that are propelled away from the head of the comet by sunlight. In imagining that radiant light could exert a force of that kind, Kepler was centuries ahead of the game.

As you are not bowled over when you step out of doors on a sunny day you could be forgiven for thinking that the pressure of sunlight is a negligible thing. Yet in the vacuum of space it becomes a force capable of driving spacecraft around the Solar System at very low cost. In the 1970s American engineers figured out a plan for sending a spacecraft to rendezvous with the comet *Halley* at this apparition, using a dozen windmill-like sails each seven kilometres long and made of aluminised plastic film only thousandths of a millimetre thick. Although this bold scheme for solar sailing was abandoned at the time, the prospect of riding on a sunbeam remains a reasonable one, for future astronauts as well as for comet dust.

Thinness is all, because only if the exposed area is large compared with the mass will the sunlight have much effect. The smaller and thus 'thinner' grains of dust in a comet are easier to shift: diminish the diameter of a particle to one tenth, and you multiply the acceleration due to sunlight by a factor of ten. Lycopodium powder, the fine-grained spores of club moss, can be dropped down a glass tube and knocked a little sideways by an intense beam of light. By the same process sunlight can drive small particles of comet dust right out of the Solar System. Typical grains of a comet tail are estimated to be, like the projected solar sail, just a few thousandths of a millimetre thick. They are now known to contain silicates, primordial material found in terrestrial rocks.

Facing: An attempt at classifying comet tails by Johann Hevelius in the seventeenth century. (Cometographia, *Royal Astronomical Society.)*

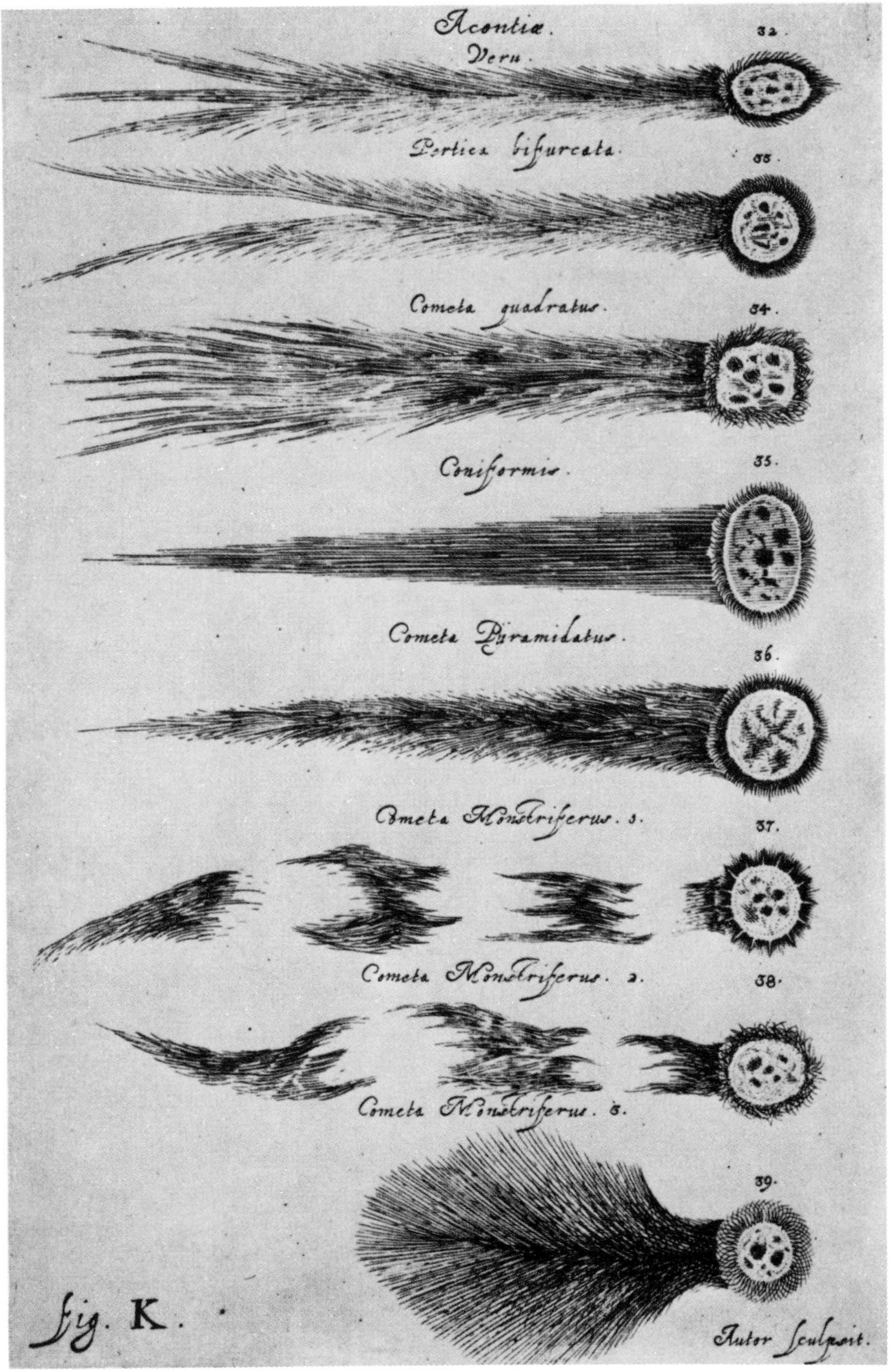
Acontiæ.
Veru.
32.
Pertica bifurcata.
33.
Cometa quadratus.
34.
Coniformis.
35.
Cometa Piramidatus.
36.
Cometa Monstriferus. 1.
37.
Cometa Monstriferus. 2.
38.
Cometa Monstriferus. 3.
39.
Fig. K.
Autor sculpsit.

The light that pushes the grains also illuminates them and we see the denser parts of the dust tail by reflected sunlight. As the dust disperses further through space it becomes invisible. The shape of the tail depends both on the movements of the comet's head and on the different accelerations away from the Sun experienced by grains of different sizes. When a comet veers sharply around the Sun the tail becomes curved like a scimitar, and perhaps scary for some Turks and Persians accustomed to military hardware of that type. Relatively large dust particles, boulders more than a tenth of a millimetre across, fan out in a sheet that is almost invisible unless it is seen edge on from the Earth. Then it makes comet lovers gasp at a fine illusion: an 'anti-tail' apparently pointing towards the Sun.

And just as a gale blows smoke ahead of a steamship that is travelling down-wind, so the pressure of sunlight carries the more easily visible dust storm ahead of a comet when it is climbing away from the Sun. But this windy analogy, apt though it is, invites confusion with the 'solar wind'. That is a true wind which produces another effect in comets and gives rise to tails of a different sort.

In the late 1940s astute astrophysicists, newly released from the British war effort, pondered the passage of the Sun and its attendant planets through the universe and concluded that the Sun should scoop up gas from the not-quite-empty space between the stars. So they boldly predicted a wind blowing into the Sun. Before long the reality of the solar wind was confirmed, although with a small modification: the wind did not blow into the Sun but out of it. So far from gaining gas from interstellar space, the Sun belches its substance into its surroundings at a rate of a million tons a day; it is the original omnidirectional fan. Comets revealed the existence of the solar wind before spacecraft recorded it directly.

Besides the tails made of dust driven by sunlight there are tails made of 'plasma' driven by the solar wind, and they stream faster and straighter from the comets. To a physicist plasma means electrified gas, in which the molecules and fragments of molecules are 'ions', carrying electric charges. There is nothing mysterious about plasma; it is the commonest stuff in the universe, and the Sun, for instance, is made of it. More suggestively for comet watchers, the familiar discharge lamps contain glowing plasma, and astronomers can recognise plasma ions in comets' tails because of the characteristic light they emit. But sunlight cannot

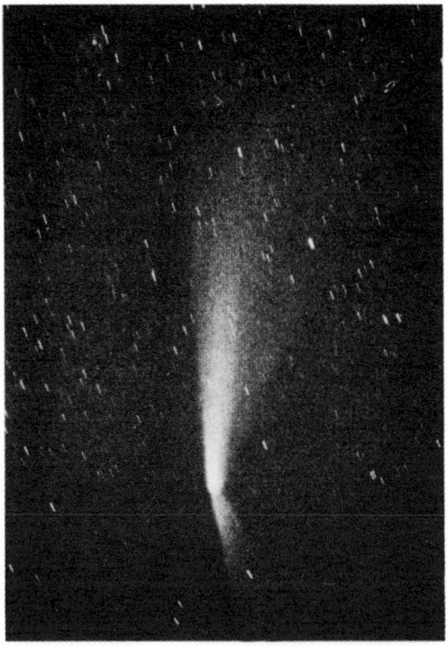

The much-remarked 'anti-tail' or 'sunward spike' of the Arend-Roland Comet 1957 (Armagh Observatory, left) was explained as a fan of large dust particles seen edge-on. The fan was shown well in a photograph taken three days earlier (R. Fogelquist, right).

Halley's Comet lost a plasma tail on 6 June 1910: the convoluted tail to the left of the main dust tail was breaking away; by next day (second photograph) the old tail was just discernible, far from the comet's head. (Lick Observatory.)

propel gas molecules in the way that it shifts dust, otherwise we should have gales every morning when the Sun rose. A different kind of pressure from the Sun was needed to fashion the plasma tails and, in 1951, Ludwig Biermann in Germany reasoned that a stream of plasma from the Sun itself could be responsible.

By that time outbreaks of aurorae, the luminous displays high in the polar atmosphere accompanied by magnetic storms and radio blackouts, were being blamed on puffs of plasma from the Sun. These events on the Earth occurred soon after bright flares appeared on the Sun's visible face. But to keep the plasma tails of comets erect at all times required, in Biermann's scheme, a non-stop flow of plasma from the Sun. A telling point for Biermann was that a plasma tail did not point precisely away from the Sun, but was just a few degrees off – exactly what you would expect if a fast-moving comet is releasing its gases into a very high-speed wind. And this reasoning was confirmed by direct investigation, when instrumented probes launched into interplanetary space detected and explored the continuous solar wind.

Blowing outwards at 400 kilometres a second, the solar wind carries with it magnetism imprinted by the Sun. When it encounters the vapours in a comet's head, electrons in the wind ionise them, making a plasma. The magnetised wind entrains the comet's plasma and sweeps it at high speed away from the Sun. Although the plasma tail is generally slimmer than the dust tail, magnetic interactions between the two wriggling plasmas, the solar wind's and the comet's, can produce curious knots and corkscrews of the tail. Such features delighted comet lovers when they appeared in *Kohoutek* in 1973–4.

The study of the plasma tails of comets borrows a little respectability from the further discovery by spacecraft that the planets, including the Earth, also have comet-like plasma tails pointing away from the Sun. Astronomers and burglars can be glad that the Earth's tail is not luminous. The solar wind produces a tail as it collides with either the outer atmosphere or the magnetic sphere of influence of a planet. Our own planet has a strong magnetic field and the Earth's bow wave, where it sweeps the solar wind aside, stands about 100,000 kilometres above the noon-time surface. The wind feeds atomic particles into radiation belts girdling the Earth. A strong gust, vented by a solar flare, can shake the Earth's magnetic defences and dump particles into the atmosphere, causing the aurorae and radio impediments. And this terrestrial analogy with the comets gives added excuses for de-

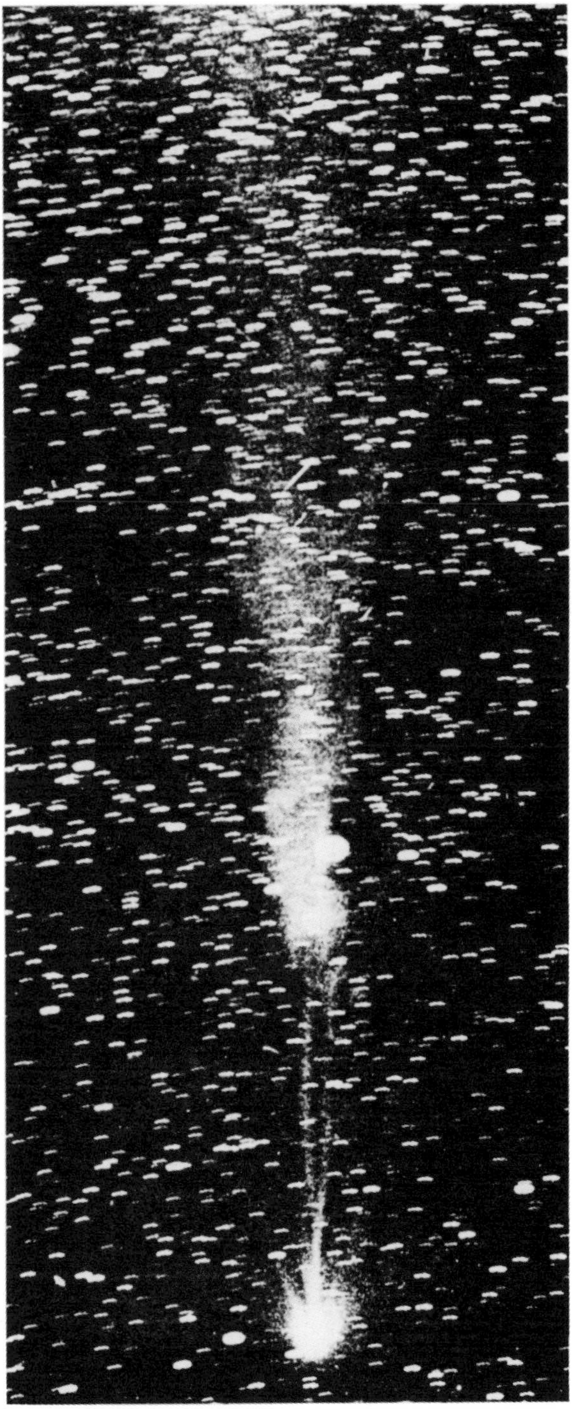

Morehouse's Comet 1908 sheds its plasma tail. The most conspicuous part of the tail detached itself and drifted away. The comet soon 'grew' a new tail. The modern explanation is that the comet encountered a region where the magnetism of the solar wind was reversed. (Yerkes Observatory.)

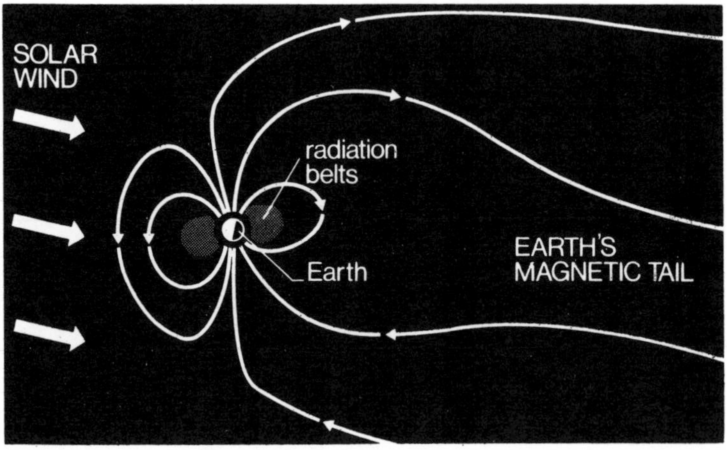

Comet-like Earth: although it is invisible, a magnetic tail exists, produced when the magnetised wind of plasma from the Sun encounters the planet's magnetism.

signing instruments that are meant to fly through a comet's tail and to examine the plasma at first hand.

Yet another of those posthumous last laughs is wafted on the solar wind. The man who found more comets than anyone else before or since, the ex-porter Jean-Louis Pons, was an unassuming man who would innocently ask for any advice that might help him in his search for comets – a habit that left him open to having his leg pulled by sophisticated scoundrels. That was certainly the intention of the Hungarian astronomer Franz von Zach. The historian of comets, Peter Lancaster Brown, relates that von Zach told Pons to look for comets whenever the Sun was covered with spots. Pons later attributed his great success as a comet hunter to this very policy, and von Zach's advice may have been far sounder than he intended. Sunspots are signs of violent magnetic activity on the Sun, which gives rise to strong gusts in the solar wind and they in turn may stimulate any comets they encounter, making them easier to detect. Efforts are in progress to relate some mysterious fluctuations in the brightness of comets to just this effect.

The character of the solar wind also explains events that startled astronomers who watched comets too diligently and would sometimes report seeing a comet shed one tail and grow another. Twentieth-century photographs confirmed this behaviour, which resembles nothing so much as a reptile sloughing off its skin. The reason for it is now known. The magnetism embedded in the plasma of the solar wind maintains the same direction (magnetic north pole versus south pole) across a wide swathe of interplanetary space, where a comet might be plying.

But as the Sun slowly spins it brings that sector of the Solar System under fire with plasma magnetised in the opposite direction. When a comet and its plasma have adjusted comfortably to the prevailing magnetism of the solar wind, encountering the zone of reversed magnetism causes electromagnetic chaos that can decouple the plasma tail. 'Extremely important and exciting,' say the comet experts.

Balls of wax make toy comets for experimentalists in Moscow, who expose them to a simulated solar wind – shots of ionised hydrogen gas travelling at an exaggerated speed of 100,000 kilometres a second. Vapour flows from the wax model and forms a plasma tail, with its own inbuilt magnetic field. The ionised matter in the tail seems to reach a speed of 10,000 kilometres a second, even within the confines of the apparatus, and the vapour in the head and tail glows just like a comet, to the joy of all concerned.

While the dust tail of a real comet is lit by reflected sunlight, the plasma tail fluoresces. That is to say, it is more like a discharge lamp, in which individual atoms or molecules become energised by radiation from the Sun and then, after a little delay, release the energy in the form of new light. The molecules impress their signatures on the light, in the form of characteristic colours that are more precisely defined as frequencies. The harsh yellow of sodium, familiar in street lighting, appears in comets too, showing that they do not lack that chemical element. But nearly all of the visible light from the plasma tail of a comet is due to molecules consisting of two carbon atoms joined together. This does not mean that carbon is the principal ingredient of comet vapour, only that the light given out by the molecule happens to match the range of the human eye particularly well.

To keep all this anatomy and physiology of the tails in perspective, remember that they are ridiculously tenuous. Seen through a comet the stars appear undiminished in brightness. The dust in the tail is so scattered that you might find one barely visible speck in a volume of empty space the size of a room. Even allowing for the invisible material, a large and active comet may leak a few hundred million tons of vapour and dust in the course of one visit to the Sun. If that sounds a lot, consider that it is much less than the weight of oil consumed by earthlings each year; yet it is strewn through volumes of space millions of times larger than the Earth and it disperses continuously in the interplanetary vacuum. The scantiest flame of burning gas is like solid steel by

A 'megalocomet' – one of several giant comet-like nebulae that point towards the centre of activity of the Gum Nebula, a region of star formation far away in space. The tail is about sixty light-years long – more than a million times larger than the greatest comets of the Solar System. It was identified on a UK Schmidt Southern Sky Survey plate by Brand & Hawarden in 1977 (Royal Observatory, Edinburgh.)

comparison and although I am not affronted, as comet lovers are, by the suggestion that a comet tail is just a prolonged cosmic fart, I do consider it an exaggeration.

The human combination of eye and brain is in one important respect better than the photographic plates used by modern astronomers: it can compensate for differences in the intensity of light more effectively. If you photograph a comet with sufficient exposure to show the tail, the bright head is drastically over-exposed and little detail can be seen in the picture. Processing techniques developed at the Anglo-Australian Observatory in New South Wales extract finer detail from the photographic plates, and in some cases the tracery that emerges in nebulous objects like comets vindicates the drawings of careful observers who relied on their eyes.

But careful photography reveals features in the sky not seen before, and helps to keep alive the old contest between perception and misperception. The 'megalocomets', or cometary globules, are an interesting case in point, which say quite different things to different astronomers looking at the same pictures. In the vicinity of very hot, newborn stars the photographs taken with the UK Schmidt telescope in Australia in the 1970s showed faintly glowing, comet-shaped features. But these objects are of quite enormous size, with tails many light-years long, while extended comets in the Solar System have a length of only ten light-minutes or so.

If you are not particularly enamoured of comets – and there are astrophysicists who stand aloof – the resemblance to comets of these huge streaks may seem just a coincidence. Sceptics offer to interpret them as mere heat shadows, where the gas between the stars happens to be shielded from the hot, invisible hurricane of a nearby newborn star. But if, like other astronomers, you fancy the idea that the very stars were built by comets, you may declare that these new-found objects are what they look like: great chieftains of the comet race, ceremoniously gathering in interstellar space to initiate new stars.

5. SNOWBALLS IN HELL

When midnight strikes on 31 December 1985 you can look for the comet *Halley* near the star Sadachiba in the constellation of Aquarius – according to Donald Yeomans of the Jet Propulsion Laboratory in California, who is *Halley's* chief keeper at this apparition. Simply to say 'The comet is coming!' seemed unadventurous, so he staked his reputation as a celestial tipster on the prediction that the comet will make its closest approach to the Sun in the afternoon (Greenwich time) of Sunday 9 February 1986. Anyone planning to turn a hopeful telescope and watch *Halley* coming in, or print invitations for a comet party, or launch a space probe to intercept the comet, has had recourse to Yeoman's tables for some years past. And he in turn has made use of electronic computers and what can be learnt from the frustrations of his predecessors.

Edmond Halley himself was content to say his comet would come back in 1758, adding the caution that an encounter with Jupiter in 1681 would have slightly enlarged its orbit and might delay it until early 1759. A dedicated Frenchman pushed analysis and sweaty arithmetic to the limit and, taking note of Saturn as well as Jupiter, said the comet would round the Sun in April 1759. Although it did so a month earlier than that, the exercise was regarded as a fine tribute to the Newtonian clockwork of the Solar System. The French were doing it again, before the next apparition in 1835; by that time they were taking Uranus and the Earth into account and the chief predictor was gratified that *Halley* was a mere four days behind the calculations. Even more thorough work was attempted in Germany, but it was not completed before the comet reappeared.

In the run-up to 1910, when the comet was due once more, the Germans fostered the numerological form of comet fever by running a sweepstake on the moment of closest approach to the Sun. British astronomers carried off the prize, after digging into the historical records for every possible clue to *Halley*'s behaviour

Facing: The anatomy of a comet was displayed in this photograph of Halley's Comet 1910. In the midst of the widely fanned dust tail was a narrow, 'twisty' plasma tail, while a bright nucleus was visible in the head. But the nucleus of a comet is now thought to be far too small to see as a solid body from the Earth. (Helwan Observatory.)

and considering the effects of Venus and Neptune. All the same, the comet was again late, by three days that time, and the aggrieved astronomers knew their calculations couldn't be that far wrong.

The comet was teasing them – how, they could not tell. One theory, advanced in the 1970s, set a large but invisible planet on an oddball orbit far beyond Neptune, just to hamper *Halley*. The favoured explanation, used by Yeomans in his reckonings for 1985–6, is that *Halley* carries a jet engine for fooling astronomers. So even if you are indifferent as to whether the comet says Hello to the Sun on Sunday or Wednesday, you may care to follow the fortunes of those who figure out the mechanical and chemical engineering of comets.

The head of a comet, also known as the coma, is generally denser and brighter than the tail. It can fill a space much larger than the Earth with visible matter and, in 1969, American and French space astronomers discovered that an invisible cloud of hydrogen gas, larger in volume than the Sun, can surround it. Yet the head is scarcely more substantial than the tail. The lack of mass in comets appears in their gravitational ineffectiveness: they do not perturb planets, or the moons of planets, even when they pass very close to them.

There seems to be a bright speck of light in the middle of the coma, and the astronomers have often suspected that a comet has a sizeable solid nucleus. In May 1910 the head of *Halley* passed between the Earth and the Sun and an American expedition sailed to Hawaii in anticipation of the event, hoping to see the nucleus of the comet in silhouette. Had it been even a thousandth of the volume of the Moon it would have shown up clearly. The astronomers might as well have stayed at home: they saw nothing at all.

This negative result gave a fillip to the admirably simple idea that the head of a comet might be just like the dusty tail only more so. Its general brightness and any brilliant nucleus were said to represent concentrations of particles akin to those let loose in the tail. This theory about the constitution of comets arose in 1870 and was dominant until 1950. Henry Norris Russell of Princeton described it confidently in 1945:

> The accepted view of the nature of comets is that they are loose swarms of separate particles, probably of very different sizes, separated by distances great in comparison with their own diameters.

The small particles were supposed to move around the Sun more or less independently and to rely on the similarity of their orbits, more than on the mutual gravity between them, to remain bunched together. The name usually attached to this picture of a comet, the flying sandbank, is unsuitable because the particles in any sandbank are close together and, if a comet were like that, the agglomerated dust grains would have little independence. Looking for a term that better conveys the idea of a loose swarm of separate particles, I prefer to call it the orbiting hailstorm.

The theory would not have hung about for eighty years, and been countenanced in popular books as late as the 1970s, had it failed to give a plausible account of some of the obvious features of comets. In 1953 Raymond Lyttleton of Cambridge refined the theory and explained how the Sun, passing through interstellar dust, assembled the hailstorm comets in its wake. Lyttleton is a bonny fighter who, despite growing isolation, has championed this theory right into the 1980s.

The 'accepted view of the nature of comets' now has them as snowballs. They do not show up against the Sun, it is said, because they are too small – just a few kilometres in diameter. That so modest a nucleus should grow a flamboyant head and tail of little substance but enormous size, is the analytical basis for regarding a comet as a swindle. The newer theory copes successfully with more and more aspects of comet behaviour, and under the barrage of snowballs the hailstorms shrink away to that limbo, somewhere beyond the Andromeda Nebula, where old astronomical theories go when they are out of favour.

Should you see a car with a Massachusetts plate bearing the word COMETS in place of the usual license number, be warned that you are sharing the road with a driver whose thoughts are millions of miles away. He is the Snowball Maker, the elderly but ever-lively Fred Whipple, who seized the throne in Cometsville half a lifetime ago. For comet lovers Whipple's worldly attainments, as the former director of the Smithsonian Astrophysical Observatory and a leading light in the early American space programme, are beside the point. Even the half-dozen comets he has to his name are less important than his portrait of a comet's nucleus as something tangible that you might fly to, stand on, scuff with your foot and use for building dirty snowmen. Voltaire would have approved: two centuries ago he told of a giant from Sirius who hitch-hiked around the Solar System by stepping on to passing comets.

Leaders of comet science with the famous Whipple 'Comets' car at the Smithsonian Astrophysical Observatory in Cambridge, Massachusetts. From left to right: Zdenek Sekanina, Fred Whipple and Brian Marsden.

Whipple nurtured the idea of the dirty snowball for several years before he published it, thinking that it could not be original. In that he was right, as we now know, because Laplace and others had written about comets in similar terms in the nineteenth century, but Whipple underestimated the extent to which later comet experts were indoctrinated with the orbiting hailstorm theory. When in 1950 he published his proposition that a comet has a definite nucleus consisting of ices intermingled with earthy material, it shook the little world of comet science. But it was a powerful theory and it explained, among other things, why the comet *Halley* had been a few days late in 1835 and 1910.

The chief exhibit in Whipple's evidence against the hailstorm was that other irritating comet, *Encke*, which had made dozens of recorded revolutions around the Sun in its diminutive orbit. It was hard to imagine, Whipple argued, that small particles in a hailstorm could endure repeated exposure to intense heat from the Sun and continue to spew out gas and dust. But if the core of the comet were a single body, a dirty great snowball, it would shed proportionately less of its total mass on each exposure to the Sun. The snowball would be incomparably larger than the hailstorm particles, yet still very much less than the glowing head of the comet.

A more subtle point about *Encke* was its unpredictable behaviour. It was tending, at each revolution, to lose some energy of motion and ease itself into a smaller and briefer orbit – a change that gravity alone could not explain. Johann Encke himself found that his comet was passing the Sun $2\frac{1}{2}$ hours too early on each three-year orbit, and he was miserable about it. By 1950, although the effect had somewhat diminished, the comet was still coming to perihelion passage up to an hour ahead of its cue. Was *Encke* losing its energy in some kind of viscous aether filling interplanetary space? That explanation was discredited when other comets, including *Halley*, were judged to be *gaining* energy of motion, which enlarged their orbits and made them appear late on cue. To provide another force besides gravity that could act on the comets, Whipple showed how to make a jet engine out of a snowball.

To understand it, imagine that you are a snowman rooted to the icy surface of the comet as it nears the Sun, and you are wondering when your head will boil off. Like the Earth, or virtually any object in the universe, your snowball is spinning about an axis and you experience alternations of day and night as your patch of the comet turns towards the Sun or into the shadow. The nights are very chilly, but at sunrise the rays begin to warm you. The snow that comprises your head does not melt; its surface layers evaporate (sublimate) directly into space. Your decomposition begins slowly in the morning, but proceeds much more rapidly later in the day when the heat absorbed at midday is reinforced by continued sunshine. In short, the surface of the comet exudes vapour at the greatest rate at siesta time.

The effect, in Whipple's theory, is a jet of vapour concentrated in the direction of the afternoon sky, as a comet dweller would

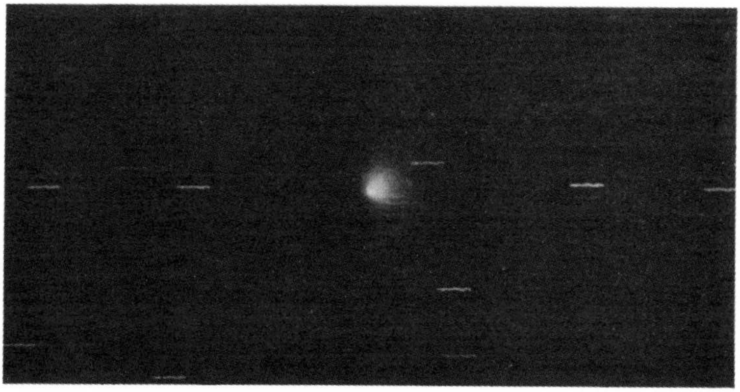

Encke's Comet, the erratic behaviour of which has figured prominently in the snowball theory of comets since its inception. It reappears every 3.3 years and has been recorded on more than fifty visits to the Sun. (G. Van Biesbroeck, Yerkes Observatory 1937.)

register it. Like the control motors carried by spacecraft, the vapour produces a thrust on the comet in the opposite direction. In part, it edges the comet almost imperceptibly away from the Sun, but in addition the thrust can act either as a brake or an accelerator. It depends on whether the afternoon boil-off occurs somewhat ahead or somewhat astern, in relation to the comet's direction of flight; that in turn is a matter of which way the comet is spinning.

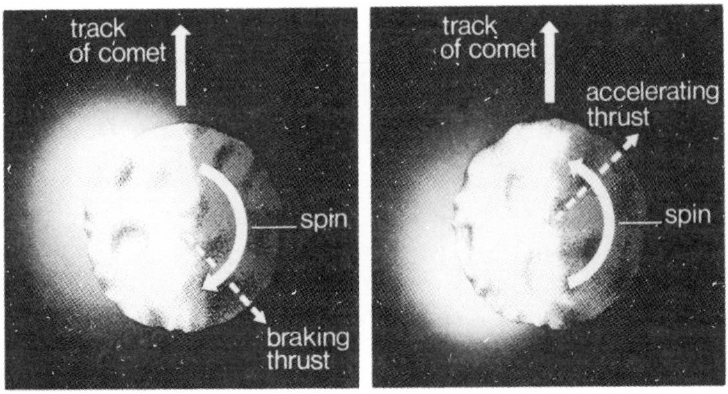

Encke spins the 'wrong' way, so that its jet engine destroys some of its energy of motion and makes it slump into a smaller orbit. *Halley* spins the 'right' way and gains energy, to the small degree that confounded the comet predictors in 1910. As the direction of spin is a toss-up, the snowball theory predicts that half of all comets should gain and half should lose energy by jet action; a census of more than thirty comets with affected orbits now shows this to be the case.

How does a Snowball Maker spend his working hours? Vault thirty years from the time when Whipple invoked *Encke* in formulating his theory, and see what Whipple and his colleagues Zdenek Sekanina and Brian Marsden had made of that same comet by 1980. After elaborate computer tests, fitting the snowball model to the facts about *Encke's* erratic behaviour during two centuries of observation, the comet seemed to be piebald.

The nucleus of *Encke*, so Whipple and his associates suggest, is a rough-hewn but almost spherical snowball, about two kilometres in diameter, and spinning around its shortest axis every $6\frac{1}{2}$ hours. One polar region is bright and active, the other very dirty, possibly coated with dust during a spell a few centuries ago when that end never saw the Sun close to. *Encke's* axis of spin gradually swivels around, like a spinning top or gyroscope that has started to wobble. The changing direction of tilt explains why the braking action of the jet engine is much less effective in our century than it was in Johann Encke's day. By 1990, the jet should become an accelerator rather than a brake. By such niceties are comet watchers sustained at four o'clock in the morning.

Spin a spotty ball and it will wink at you rhythmically, like a lighthouse. For example, a pulsar is thought to be a collapsed, spinning star with bright patches created by the magnetic poles. And the pulsating comet *Donati* of 1858, while awkward for the hailstorm version of comets, became easy game for snowballers. *Donati* threw off a succession of bright haloes, expanding one after the other on the sunward side. The hailstorm explanation required converging swarms of particles that struck one another like flints and released vapour and fine dust. That has been tested in a wind-tunnel and, somewhat surprisingly, it works, but it does not account for the regularity with which the haloes appeared, at intervals that were always multiples of 4.6 hours. In Whipple's words, '*Donati* ran like clockwork for three weeks'. If the core of *Donati* is a snowball that spins on its axis every 4.6 hours, and on each rotation it exposes one very volatile spot to the Sun's rays, the puzzle is solved.

For the break-up of comets, too, into two or more separate heads, the snowball theory gives readier answers than the hailstorm theory. While it is easy to imagine an orbiting hailstorm being thoroughly disrupted and dispersed by tidal forces and the pressure of sunlight, it seems difficult to form separate but coherent heads. For a spinning snowball, on the other hand, it is entirely natural that centrifugal forces, the explosive release of gases or the tidal effects of the Sun should sometimes tear it apart. The brightening that often accompanies a break-up is explained by the exposure of fresh surfaces of volatile material to the heat of the Sun.

Any attempt to define the snowball called *Halley* relies heavily on observations made in 1910 with instruments far inferior to those of today, and on comparisons with other comets. To help people who contemplate sending spacecraft to meet the comet at this apparition, Ray Newburn of the Jet Propulsion Laboratory assembled a description and *modus operandi*; I take his 'nominal' figures, although for most items in the specification he offers a wide range of possibilities. *Halley*, then, is tentatively portrayed as a snowball five kilometres in diameter, weighing at least

65 billion tons and spinning on its axis every 10.3 hours. Of this mass, forty per cent is water ice, ten per cent other volatile substances, and fifty per cent 'solid' material, dust and stone. These constituents are said to be well mixed, so that when vapours come off in the heat of the Sun they drag dust with them. Gravity is so weak at the surface of the snowball that if you tossed a penny upwards at only two metres per second you would not get it back. On the basis of these and other assumptions, Newburn will tell you, for instance, that if you fly a spacecraft a thousand kilometres from the nucleus on a certain date you should expect it to be shotblasted by a hundred million grains of dust, but none of them larger than a millimetre.

Steam-engine enthusiasts may have noticed already that the chief outpourings of a comet are steam from the vaporised ice mixed with a thick smoke of dust. But no sooner has it done its work of jet propulsion than the water vapour is split by the Sun's ultraviolet rays into its component atoms. The astronomers' spectroscopes detect little water in the comet's head. The silicate grains of dust escape dismemberment, except in sungrazing comets, but other volatile molecules from any comet suffer the same prompt degradation as the water. Some of the chemical fragments react together to produce new compounds. All in all, the composition of the head and tail of a comet bears little resemblance to the internal chemistry of the snowball.

After the virgin comet *Kohoutek* was scrutinised by radio and by visible, ultraviolet and infrared light, one could sum up the detectable constituents of comets in general in one word: noisome. Poisonous ingredients such as hydrogen cyanide (prussic

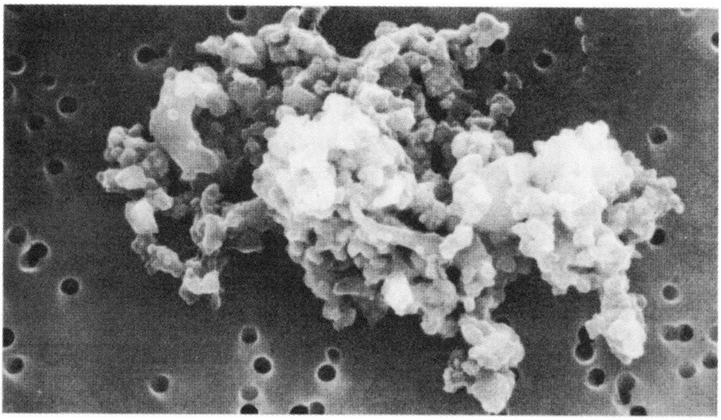

Comet dust: small extra-terrestrial particles collected by a NASA U-2 aircraft flying at high altitude, and shown magnified about ten-thousandfold. Constituents include silicon, magnesium, iron and carbon. (D. Brownlee, University of Washington.)

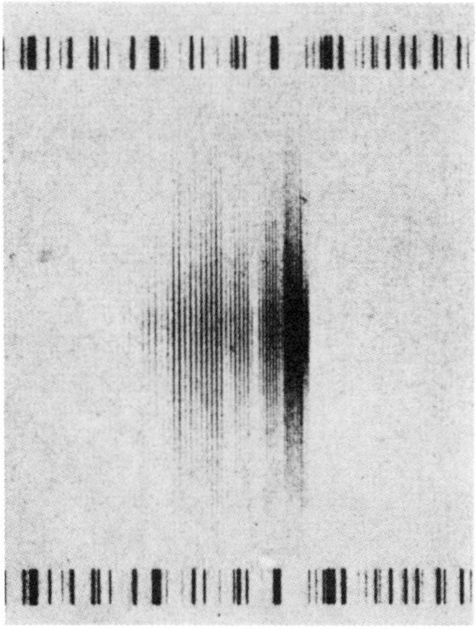

A poisonous gas shows up in a comet. Evidence for the presence of particular materials comes mainly from the light of characteristic frequencies (or wavelengths) that they emit. In this case, the 'fingerprint' of cyanogen appears plainly in a spectrum of light from Kohoutek's Comet, analysed in January 1974. The discovery of cyanogen in comets was responsible for much alarm when the Earth passed through the tail of Halley's Comet in 1910. (Observatoire de Haute Provence, CNRS.)

acid), methyl cyanide and carbon monoxide are well to the fore, amid hydroxyl radicals (OH) and various other combinations of hydrogen, carbon, nitrogen, oxygen and sulphur atoms. At least nine metallic elements – sodium, iron, copper and so on – have been identified. Reasoning back from these known products, Armand Delsemme of Toledo suspects that perhaps six per cent of a comet's mass consists of frozen carbon dioxide and carbon monoxide, and another one per cent of cyanides and more complicated compounds of carbon. So far, alcohol has not been detected in any comet, but I shall be surprised if it fails to show up in *Halley*.

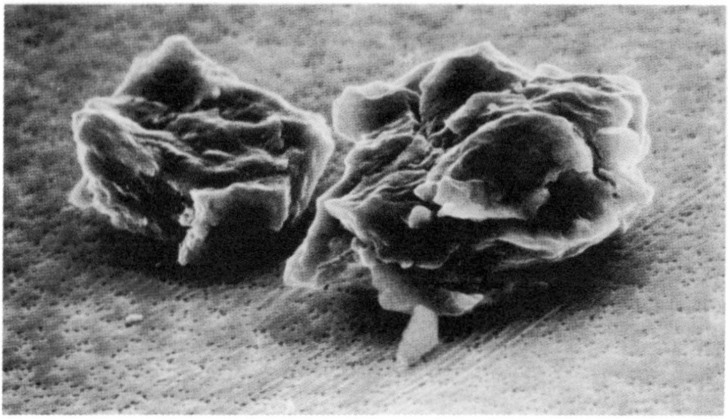

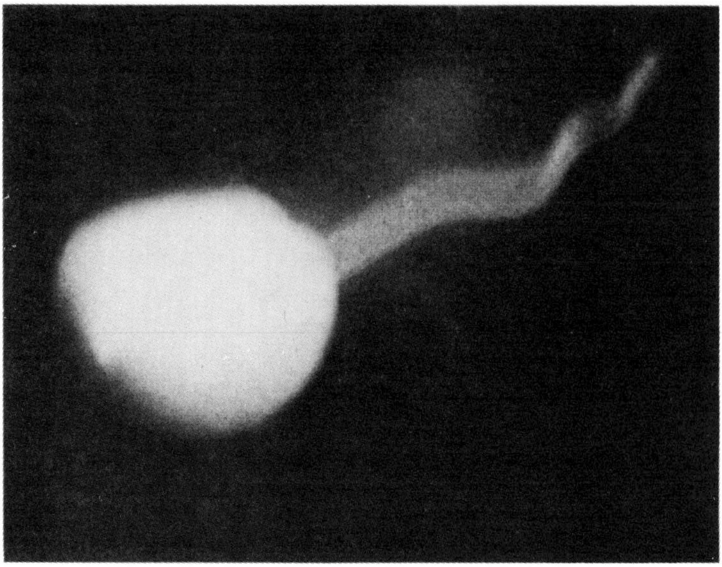

The fireball of a meteorite that fell in New Mexico in March 1933, captured by a quick-witted amateur photographer.

There is plenty of alcohol in interstellar space. Since the 1960s radio and infrared telescopes have identified more than forty different chemicals in the clouds of gas and dust, including many organic, carbon-based compounds that earthlings generally associate with life and death: alcohol, formic acid, formaldehyde and so on. But these are products of random chemistry, on a par with the primeval ice and stone that built the planets around the Sun. Even if, as some say, the interstellar dust might include grains of cellulose, as in wood, we need not assume that any alien species is running a sawmill out there. Comets cannot have failed to take on board alcohol and all sorts of other materials during their formation, and they probably replenished their supplies during later encounters with interstellar clouds.

The comets can therefore be thought of as messengers from the past, conveying materials from the primeval dust-disk of the Solar System and from interstellar clouds, into the realm of the inner planets – and perhaps on to the surfaces of the planets. One suggestion is that the Earth acquired its oceans and early atmosphere by multiple impacts during the Age of Comets, rather than from volcanoes as a more straightforward theory has supposed. If that were so, a generous supply of carbon compounds would have come at the same time, and perhaps played a part in contributing raw materials from which life could originate. It is difficult, so long after the event, to identify the cometary role in the chemistry of the early Earth, but there are less direct, and continuing, methods of transferring material from comets to our planet.

Even while *Halley* was far away in its 76-year tour of the Solar System it left a memento in the form of twice-yearly showers of meteors, or shooting stars. A comet's tail strews the orbit with dust grains like confetti after a wedding, and these continue to circulate, albeit invisibly. Eventually they become spread all around the orbit, so that every year in October and early May, when the Earth passes close to *Halley's* orbit, the larger dust grains come raining down, streaking brightly across the night sky. A piece the size of a pinhead puts on quite a show, as it 'burns up' in the friction of the Earth's outer atmosphere, a hundred kilometres high.

Other annual meteor showers, famous among astronomers, correspond with the orbits of other comets, known and extinct, and radars can detect meteor showers even when the Sun is up: for instance, a regular daylight shower at the end of June is associated with the comet *Encke*. But the dust disperses continually and is subject to the same kind of gravitational football from the planets as comets are. In time, the waifs become impossible to identify with their parent comets, and they can encounter the Earth singly at any time. Hundreds of tons of meteoritic material settle unobtrusively on the Earth every year.

The smallest meteorite particles, probably more typical of the contents of comets' tails, do not become incandescent when they hit the Earth's atmosphere, but descend quite gently. Donald Brownlee of Seattle has used high-flying U-2 aircraft to collect particles of dust that are believed to be minute pieces of comets. These micrometeorites are typically a hundredth of a millimetre wide and under a microscope they look like fish-roe. Similar material has been raked by magnet from the ocean floor, to where it evidently descends. The composition of Brownlee's particles is similar to that of the most primitive objects known to science, the carbon-rich stony meteorites.

Larger chunks of interplanetary debris reach the ground as meteorites because they fail to burn up completely in the atmosphere even though they come in very fast. Perhaps a hundred tons a year arrive at the surface, and a few hundred grams are collected for museums. In 1794, the distinguished French chemist Antoine Lavoisier wrote a report saying that meteorites could not possibly come from outside the Earth; he was promptly beheaded by the revolutionaries' guillotine, although not for that reason.

The investigation of sky pollution began to broaden out from purely cometary studies when, at the end of the eighteenth century, scientists accepted that stones really did fall out of cosmic space.

The stones themselves had, of course, been known from prehistory and our forefathers were acquainted with iron in its native metallic form, in a certain class of meteorites. But the very name 'meteorite', with the same root as meteorology, implied an origin in the atmosphere rather than the universe. Between 1794 and 1803 a succession of well-observed falls near Siena (Italy), Scarborough (England) and l'Aigle (France) persuaded the scientists of Europe that the stones were extraterrestrial. In 1798, students at Göttingen established that meteors, too, came from above the atmosphere.

By coincidence, in 1801 Giuseppe Piazzi of Palermo spotted a new kind of object far away in the Solar System. On the principle of 'if it moves, call it a comet', he first thought he had discovered a comet without a tail; but it proved to be a small planet. Ceres, Piazzi called it, and it was the first and largest of many minor planets, classed as 'asteroids' by William Herschel, that were found to be orbiting the Sun in a broad belt between Mars and Jupiter. Nowadays the asteroids are thought to range from a thousand kilometres in diameter, down to lumps of rock the size of mountains, or even fists. In aggregate their mass is about one two-thousandth of the Earth's.

Thus the known varieties of litter in the Solar System quadrupled in a few years around 1800: to comets were added meteors, meteorites and asteroids. That comets were the source of meteors became clear in the mid-nineteenth century, but there was no obvious link with meteorites until 1932 when Karl Reinmuth of Heidelberg found another kind of minor planet – not safely parked away beyond Mars but crossing the orbit of the Earth. It was named Apollo and it was the first of a new class of asteroids, more conveniently called 'apollos' to distinguish them clearly from the ordinary asteroids of the main belt between Mars and Jupiter. The only objection to the term is that it implies a self-centred view, because this class of objects really includes the 'amors' which come closer than Mars without yet reaching the orbit of the Earth. While the main-belt asteroids generally have female names the apollos, like their prototype, are mostly male, although this discrimination has ceased in recent namings.

One theory saw them as ordinary asteroids, or fragments,

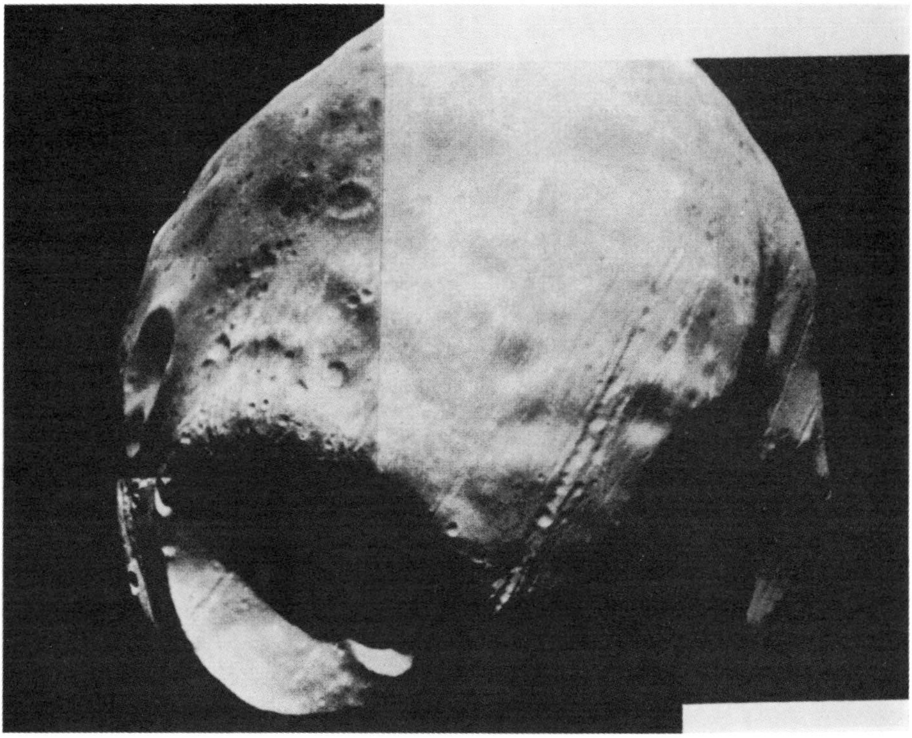

Phobos, one of the micro-moons of the planet Mars, gives an impression of what other small bodies of the Solar System may look like – whether main-belt asteroids, apollos or the nuclei of comets. (Viking Orbiter I mosaic, 1978.)

dislodged from their normal orbits beyond Mars. But collisions with planets, and gravitational football booting the apollos out of the Solar System, clear them away continuously and the main-belt asteroids seem incapable of maintaining the supply. Comets can easily account for them, though, if the apollos are mainly comets that have run out of steam. There are counter-arguments: the statistics may well be biased by the limited scope of observations of small objects in the Solar System, and comets might in any case disintegrate completely into dust and vapour. On the other hand, we may be handling substantial chunks of comets already.

Some of the well-known meteorites, far smaller than apollos, that are housed in the world's museums may be pieces of comets and apollos. The strongest candidates are the primitive stony meteorites rich in carbon and carbon compounds. Astronomers see comets spewing out fragments large and small, and they visualise both comets and apollos making collisions with ordinary asteroids in the main belt. By these processes, meteorites of all sizes and types might be produced, but because many meteorites have evidently been melted, long ago, a view persists among

In the carbon-rich Allende meteorite, the white lumps may represent some of the oldest and most primitive solid material that went into the making of the Solar System. Peculiarities of atomic composition suggest that they were from the debris of an exploding star, and they may have provided a special source of radio-active heating for meteorites and comets.

specialists that the source must have been a large asteroid or small planet, many kilometres at least in diameter, which broke up. For present purposes it is enough to suppose that *some* meteorites originated from comets, so that comets contain fragile stony aggregates. These they contribute, along with meteor dust, to the general pollution of the inner Solar System.

To see comets in this light focuses attention in a practical fashion on the ways in which the pollution might physically injure the inhabitants of the Earth. The principal suggestions for examination are those that hold comets to be harmful to health, (1) because their meteor swarms contain dangerous germs, and (2) because large impacts by comets or apollos derived from comets can kill most life on Earth. Putting comets in their place in the litter-cycle is also the best remedy for the superstitious forms of comet fever. No one need mistake an ice-encrusted junk heap for a personal message from God.

There are enough recipes for comet snowballs to stock an ice-cream parlour, and in that respect the theory is perhaps too accommodating. The brightness of comets suggests to Delsemme that typical 'new' snowballs are three kilometres in diameter, while the largest might be several hundred times more massive and perhaps twenty kilometres across. Other astronomers suspect that the comet of 1729 may have been more than 100 kilometres in diameter. The only essential ingredients for the 'dirty snowball' (or 'snowy dirtball') are plenty of water ice and stone dust, plus some flavouring with compounds of carbon, nitrogen and so on. After that the imagination of the confectioner is unrestrained. He can have a consistent mixture, ice and dust all the way through, or like a delinquent schoolboy he may hide a big stone inside his snowball. Or he is free to scatter small stones through the mixture, in the manner of angelica and chopped walnuts in a cassata. Then he may wish to decorate the exterior with humps, cracks and craters, and put secret pockets of gas or highly volatile materials under the surface.

Are the snowballs strawberry-coloured? The stony material in the comet nucleus may be painted a very dull red, according to planetary astronomers at Cornell University, because that is the colour of the so-called trojan asteroids, small objects with homeric names grouped at set intervals around the orbit of Jupiter. These astronomers think that the trojans are closely related to the comets. Their colour is well mimicked by the waxy carbon compounds known as kerogens, which are found in coal tar on the Earth and also in some meteorites. Mix them with ice and it seems to me that the result might be a sickly pink.

In the absence of any firm rules, the method of assembly is also a matter of taste. Weak gravity, vacuum welding and chemical bonding are the chief ways of holding the ingredients together. In view of the propensity of comets to break up, a loose assembly of several small snowballs is not disallowed. Some theorists make their comet at the birth of the Solar System and leave them severely alone in the freezer of the Öoo Cloud until the time comes to serve them up to spectators near the Sun. Others repeatedly respray their snowballs with ice, dust and flavouring, whenever the Sun passes through an interstellar dust-cloud.

For those who prefer their snowballs cooked, there are several possible sources of heat: from collisions between comets, or between comets and dust grains, which warm them without entirely smashing them; from chemical reactions between the components of the flavouring; and from radioactivity in the dust and stones. To pursue this last and most interesting possibility for a moment: a large carbon-rich meteorite fell in Mexico in 1969 and was found to contain white lumps up to a centimetre wide, consisting of minerals rich in calcium, aluminium and titanium – just what the experts would have expected from the first cooling of a cloud of dust and gas around the newborn Sun. But during the early 1970s the white material turned out to have peculiarities in its atomic composition, which inspired the new story that the Solar System was formed shortly after the explosion of a nearby star. In this narrative the white lumps are now said to be products of the waves of gas racing out from the exploding star as, in its death throes, it manufactured fresh supplies of the elements needed for building planets, comets, meteorites and eventually living things. There were extra doses of radioactivity as well, notably in radioactive aluminium, which greatly augmented the sources of internal heat available for melting materials in quite small objects, comets for instance.

Any cooking is generally supposed to have ceased a very long time ago and water that may have melted inside the comet is deemed to have refrozen. Conceivably nature follows every recipe, in one comet or another, and none of these variations affects the most obvious behaviour of live comets. But the choices relate, as we shall see, to the question of whether smallpox comes from a comet.

In summary, Whipple's theory of comets despatches snowballs inwards from the coolness of the Öoo Cloud into the inferno close to the Sun. There they suffer decay and disintegration. They lose dust into their tails, making meteors and micrometeorites. They may break up, making multiple heads, and also throw off smaller fragments that seem to be the source of some types of meteorites. If a comet is detained in the inner Solar System, it will expend its ices and cease to be a comet, but it may continue to orbit the Sun as a somewhat menacing agglomerate of other material – the remains of the filling of the snowball.

Could it all be a myth? A starlike speck in the midst of a glowing comet's head, or a distant view of the inactive comet far from the Sun, are the best that powerful telescopes can show anyone of a comet nucleus. Raymond Lyttleton, the dogged defender of the orbiting hailstorm, spoke of the snowball theory in 1976 in terms that connoisseurs of cometary rhetoric will appreciate:

> If the claims of science and reason are waived, there could be said to be something heroic about the way in which for a quarter of a century the numerous icy-nucleus proponents have stuck to their guns with not a shred of acceptable scientific evidence to support the theory. . . . . An effusion of papers that mainly talk their way through the alleged physical processes describe and even offer sketches of purely fanciful home-made models of the micro-nucleus.

For all his hyperbole Lyttleton has a point. Some respected theories now deceased were equally elaborate and promising; one thinks of those confident diagrams of mountain building prevalent before the 1960s, and how the learned geologists blushed when continental drift was at last confirmed. Not many years ago plenty of astronomers believed there were active volcanoes on the Moon and living things on Mars.

The only way to make certain of the snowball is to send a spacecraft to a comet to take close-up pictures of its nucleus or to land on it. Best of all, the craft might land and take off again, returning to the earth with frozen samples of pristine snow, to enable scientists to analyse the most primitive form of the material from which planets were built. Planning a mission involves an argument of elegant circularity. You want to find out if the comet is a snowball, but to design the spacecraft and its programme of work you have to know what a comet is like, so you assume it is a snowball. Such a procedure, quite common in science, is known technically as picking yourself up by your own bootstraps. And among those few who still wonder if the hailstorm version might not be correct after all are space engineers worried about the shotblasting of a comet probe. If the snowball theory were wrong, they fear that an expensive mission might be wiped out in a moment, by a volley of hailstones travelling at 200,000 kilometres an hour.

If a well-equipped spacecraft should fail (against all expectation, I should say) to find a snowball in the core of a comet, the hailstorm theorists might still have the last laugh. Only the experts have much reason to care whether a comet is a hailstorm, a snowball, or candy floss, but the spirit in which the answer to such a question is sought is of wider concern. However amusing and persuasive a hypothesis may be, it can be taken seriously only if it has stood up to severe tests of observation and experiment. Even then it may be punctured later, as happened to one of the greatest of them all, Newton's treatment of gravity. There is a continuous shuttle of theories, to and from that limbo of discarded concepts.

Cometology has been a fertile field for both superstitious and scientific imaginations because the objects are spectacular, which makes them seem important, while solid facts about them have been very scarce, owing to their remote, flimsy and transient nature. It is therefore a fascinating model for the interplay of argument and evidence in weightier subjects, both in science and in public affairs. Men kill one another in the name of theories of social behaviour, for which evidence is even scantier than with comets. The methods of the KGB's psychiatric services are not yet used to treat dissident opinions about comets, but anyone who thinks such a possibility ridiculous has already forgotten the fate of Galileo, and of the Soviet geneticists during the recent Lysenko era.

In the need to tolerate the most perverse ideas, there is an obvious congruence between good science and political freedom; also in the deference, by the opinionated people themselves, to verdicts delivered from the outside, by nature on the one hand and by voters on the other. But scientific research should not be mistaken for easy-going liberalism. At least in the 'exact' sciences, of which astronomy was the first, the treatment of ideas can be quite savage. Inexact theories, however fashionable, may suffer sudden death from new facts, or from new theories that fit old facts better: it can be as clear-cut as finding that Sunday is really Saturday. A Nobel prizewinner whose interests lie in the nucleus of atoms rather than in the nucleus of comets once remarked to me that theoretical physicists like himself had a moral ascendancy over all other scholars: 'Every one of us has at some time or another been proved completely and incontrovertibly wrong.'

But triumphant success is also possible, and the first predicted reappearance of Halley's Comet in 1758 was such a moment in this field. The next may be the first close-up picture of a comet snowball, emerging on a television screen at mission control. And the impatience of the protagonists to gamble all their preconceptions on a definitive test is the hallmark of healthy science. If comets were still beyond human reach, one might well take the snowball on trust, as with current theories about the core of the Sun. But a mission is technologically possible and all bets are off until it succeeds, even if that should take a hundred years.

6. AN INTERPLANETARY 'FLU MACHINE?

Some twenty years ago, when the ransacking of meteorites for symptoms of life was in full swing, encouraged by the hypothesis that a once-living planet had broken up, some small, strikingly lifelike grains turned up in a meteorite. They caused considerable excitement until a botanist spoiled the fun by identifying some of them as ragweed pollen. The investigators had either to claim that nature reinvented ragweed on another planet, or admit that their samples had somehow become contaminated.

Soon after that, an unknown hoaxer of Gascony, who had set out to mock the great men of the nineteenth century, obtruded into twentieth-century studies of life in outer space. When a famous meteorite fell at Orgueil in southern France in May 1864, someone took the trouble to glue lumps of coal and pieces of reed into one of the carbon-rich fragments. His purpose may have been to confound Louis Pasteur, who was then busy disproving the 'spontaneous generation' of life. Or perhaps the target was an eminent chemist who taught that carbon-rich meteorites represented extraterrestrial humus.

Where joking ends and faking begins is sometimes hard to tell. In one detected fake in biology, serious in intent, toads were injected with indian ink to prove the perfectibility of man; in another, intelligence-test results and even a non-existent collaborator were invented, to demonstrate that clever people inherited their talents from their parents. Honesty, it seems, was not a required factor either in perfection or in intelligence. On the other hand Piltdown Man, the fossil forgery that bemused anthropologists for forty years, may have been a practical joke that got out of hand, rather than a serious attempt to show that Adam was an Englishman.

Whatever the intentions of the French hoaxer, his plot misfired, because the pieces of the Orgueil meteorite were so abundant that no one bothered to examine his *chef d'oeuvre* for a hundred years. Then it came into the hands of Edward Anders and his colleagues in Chicago. They analysed the contents, down to the nineteenth-century French glue, and disclosed the hoax in 1964. Although modern scientists were not to be fooled by it, those working in a contentious area of research gained little comfort from this tacky farce.

The idea that life first came to the Earth from outer space has always been, in a sense, as good as any other, and more romantic than home cooking. Sir William Thomson a hundred years ago, Svante Arrhenius early in this century and Leslie Orgel and Francis Crick more recently – these are a few of the scientists who have taken the notion seriously. Because there is no natural record of events, the origin of life is a subject for informed speculation and, at best, laboratory experiments. The oldest substantial fossils, ones that you can kick as opposed to specks that you peer at doubtfully with a microscope, are reefs of algae reported from Western Australia in 1980, dating from about 3500 million years ago. The first billion years of the Earth are virtually beyond direct recall.

But take a soup of chemicals and stew it gently for a long time with a supply of energy – radioactivity, ultraviolet rays, volcanic heat, lightning strokes, meteorite impacts, you name it – and it may eventually produce living systems, less by chance than by a prebiotic natural selection of genes and proteins. Molecular biologists have a technical grasp of that story and no deep doubts about it. For most of them the Earth's surface seems the obvious and congenial place for it to happen, but the possibility that the young Earth was impregnated with interstellar or interplanetary seeds cannot be ruled out.

Comets have the water and chemical nutrients essential for life and one hint of their potential as vehicles of life is the present suggestion that *we* might seed *them*. Freeman Dyson of Princeton would like to see genetic engineers developing trees capable of growing on carbon-rich asteroids and on comets. They would be adapted to living in the vacuum of space. The comets of the Öoo Cloud give a combined surface area far larger than the Earth's, but the Sun is remote and feeble so the trees would have to be immense, many kilometres wide, to gather enough sunlight.

Seen from afar a comet tree would look like a large bush in a small pot. Other plants, together with animals and people, can live in the branches of the giant trees, bringing human evolution full circle from the jungle trees of our ancestral primates to the tree-dwellings of spacemen. If comets permeate the spaces between the stars, the gigantic vegetation can then spread quite readily from one sun to the next; if not, the seeds of the giant trees will just have to be that much hardier, to survive long journeys drifting through space, like coconuts floating from island to island, or else be transported by human star-travellers. These are some of Dyson's prescriptions for the Greening of the Galaxy.

But I come now to a set of ideas compared with which Dyson's proposal is just a cautious extrapolation of current technology. At the end of the 1970s a pair of astrophysicists in Britain, Sir Fred Hoyle and Chandra Wickramasinghe, punched protobiology in the midriff and medical science on the chin. Life on Earth, they say, began in comets, and diseases still come from them. Wickramasinghe is head of astronomy at Cardiff and a leading authority on the composition of interstellar dust. He has worked on and off since 1962 with Hoyle, a hardened pugilist, who is bloodied but unbowed from battles in cosmology and Cambridge politics, a theoretical astronomer of high talent, and a writer of imaginative fiction in which the scientist-hero is always right.

Hoyle has not attempted to disguise his comet fever; indeed, he admits to contracting it:

> Perhaps no astronomical object captures popular attention more strongly than comets, and after seeing two brilliant ones myself, both in the year 1957, comets *Arend-Roland* and *Mrkos*, I can understand why.

Within a few years of writing those words in 1975, Hoyle was demonstrating with Wickramasinghe that, whatever else comets may do, they still evoke astounding new theories.

The comets of Hoyle and Wickramasinghe are snowballs coated with interstellar food and then partly melted. At the origin of the Solar System, in the swarm of comets, each snowball gathered carbon-based chemicals and other essentials of life on to its surface, from the dust-clouds between the stars. Cellulose, the commonest material of earthly life, is said to be a major component of the dust, and they picture the comets acquiring a mantle of interstellar material, perhaps a kilometre thick.

Everything was frozen to begin with, but then melting occurred. In first formulating their ideas, Hoyle and Wickramasinghe relied on collisions between comets to spark off chemical reactions between the ingredients. These could have produced warm, watery inclusions that might survive for millions of years. By 1980 an alternative source of heat was being considered: radioactive aluminium in the main body of the comet, assimilated from the star that exploded at the time of the birth of the Solar System, would be capable of melting the core of a large snowball.

Charles Darwin visualised life beginning in 'some warm little pond' containing the necessary chemicals. Hoyle and Wickramasinghe offer enclosed ponds larger than cathedrals, in a billion comets, each of them insulated from the outside universe by hundreds of metres of frozen crust. There, viruses and bacteria would evolve spontaneously, they say. In time the igloo-ponds would freeze, preserving the viruses and bacteria in a state of suspended animation.

Around 4000 million years ago, one fertile comet brought life to the Earth, according to this theory. It may have come by direct impact, or else via the meteoritic dust from the comet's tail. Plunging near to the Sun the comet shed its outer layers like a seed pod, releasing the fragile cells into space. Most of the cells died but some, perhaps frozen and surrounded by protective ice, made a chance encounter with the Earth's atmosphere and fell to the surface. There they may have found, from previous cometary encounters, 'a supply of the chemical foods to which they were accustomed'.

If that inoculation did not 'take', if the cells died in their new environment or failed to evolve the means of living by sunlight, there was always another comet to come along, for a further attempt. And billions of years after life became established on Earth, the bipeds that eventually evolved gazed on the ancestral comets and feared them, as purveyors of disease.

At the end of 1664, when Edmond Halley was eight years old, a comet appeared and in the following year London experienced a disastrous outbreak of bubonic plague that killed one person in every five. In his *Journal of the Plague Year* Daniel Defoe, creator of Robinson Crusoe, was in no doubt that the slow motion of the comet across the sky portended '. . . a heavy judgement, slow but severe, terrible and frightful, as was the Plague'. That was one of the last attempts to link comets with pestilence for a long while, by serious writers as opposed to cranks. Newton and Halley subverted any medical theory of comets by explaining the nature of their orbits and certifying their great distance from the Earth, so that the ability of comets either to prophesy or to cause disease became doubtful.

Medicine otherwise made slower progress but, in the nineteenth century, Pasteur and others established that microbes

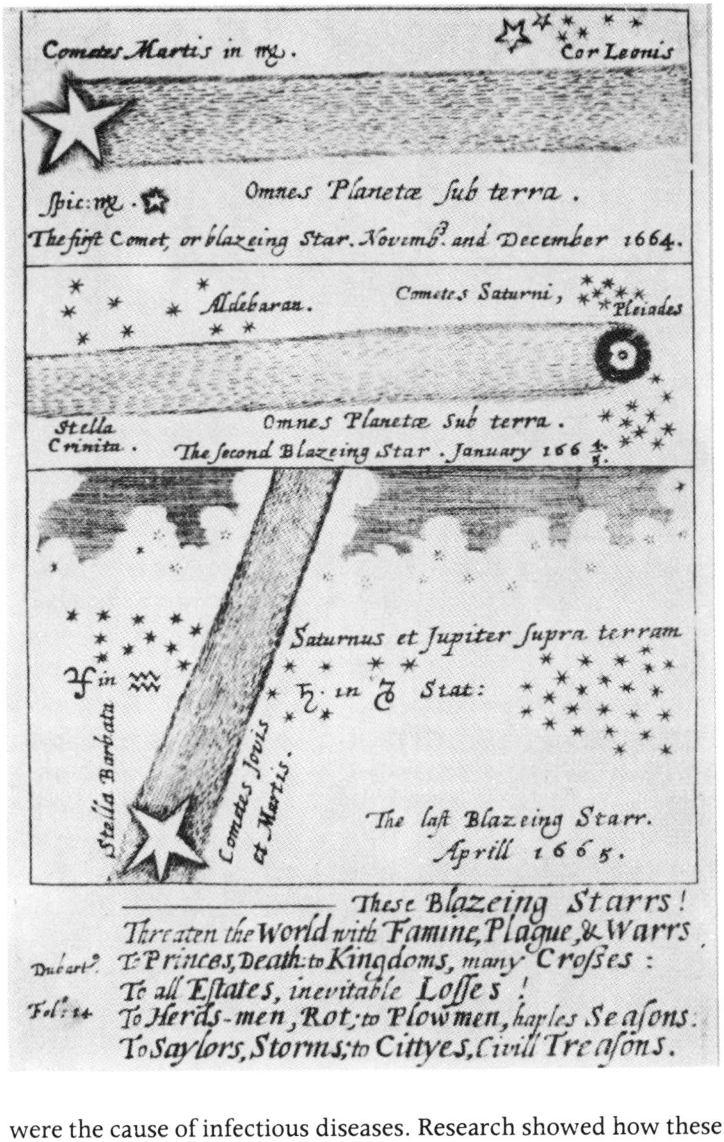

A contemporary prophecy that the comet of 1664 would bring the Plague, which did ravage London in 1665. (John Gadbury, De Cometis *1665; Royal Astronomical Society.)*

were the cause of infectious diseases. Research showed how these pathogens were spread, for example by polluted water (cholera), fleas (plague) and mosquitoes (malaria). In the twentieth century epidemiology progressed with the study of the spread of diseases among animals which can harbour bacteria, fungi and viruses that affect human populations. Military laboratories cultivated natural and modified pathogens and studied how to scatter them through the air across wide areas, as a weapon of war. And even though people in tropical areas unable to afford clean water,

Sir Fred Hoyle, the astrophysicist whose theory fills the sky with germs.

vaccines or antibiotics still perished by the million from infections, they could die secure in the knowledge that the only reason was human negligence; devils, witches and comets were not to blame.

The divorce between bodily disease and comets seemed final and absolute – until Hoyle and Wickramasinghe offered to set medicine back a few centuries by remarrying them. As the two astrophysicists portray diseases from outer space, they are a straightforward extension, and a daily confirmation, of the origin of life in comets. If comets could seed our planet with viruses and bacteria four billion years ago, why not now? If a millionth part of the meteoritic debris falling to the Earth from comets consists of viruses, a small garden could collect millions of viruses every day, ready to assail plants, pets and humans. Hoyle and Wickramasinghe imagine the various diseases that afflict mankind originating in different comets.

They are not suggesting anything as vulgar as a direct link between sightings of comets and outbreaks of disease. The danger they perceive lies in the invisible clouds of dust spewing out of comets, orbiting through the inner Solar System, and being scattered and smeared through space in very complex ways. When the Earth runs into a cloud, the germs that survive the encounter settle to the ground only slowly, while weather patterns spread them and re-clump them around the world. Once they are well established on the Earth, some diseases need no replenishment from space: tuberculosis is one. Others, polio and cholera for instance, persist of their own accord but are 'topped up' by fresh deliveries of germs from space. Finally, in this unwholesome catalogue, there are diseases like bubonic plague and smallpox that come and go as the orbiting clouds decree. Hoyle and Wickramasinghe affirm: 'We did not arrive at our ideas at all lightheartedly.'

That I believe: I have never seen Fred Hoyle in a merry mood, and who would want to jest about diseases – even influenza, which he singles out for special attention? To illustrate the patchy, sudden and widespread outbreaks that the theory requires, he cites the fatally virulent strain of influenza that first struck at human beings in Massachusetts in September 1918 and, within a couple of days, felled others in Bombay. That was before the days of intercontinental air travel, the advent of which, according to Hoyle, has had little effect on the rate of spread of influenza. Epidemiologists would explain the near-simultaneous outbreaks in the USA and India by viruses harboured among animals and spread in advance around the world by birds and domestic animals. But Hoyle scoffs at such mundane explanations: influenza, he says, appears to exist in humans only to the extent that it is 'driven from the outside'.

The old British custom of isolating children in boarding schools provided a special opportunity for testing the hypothesis that diseases come from comets. Hoyle and Wickramasinghe were not content to comb the medical literature for mysteries that they might turn to their advantage. Like good scientists they for-

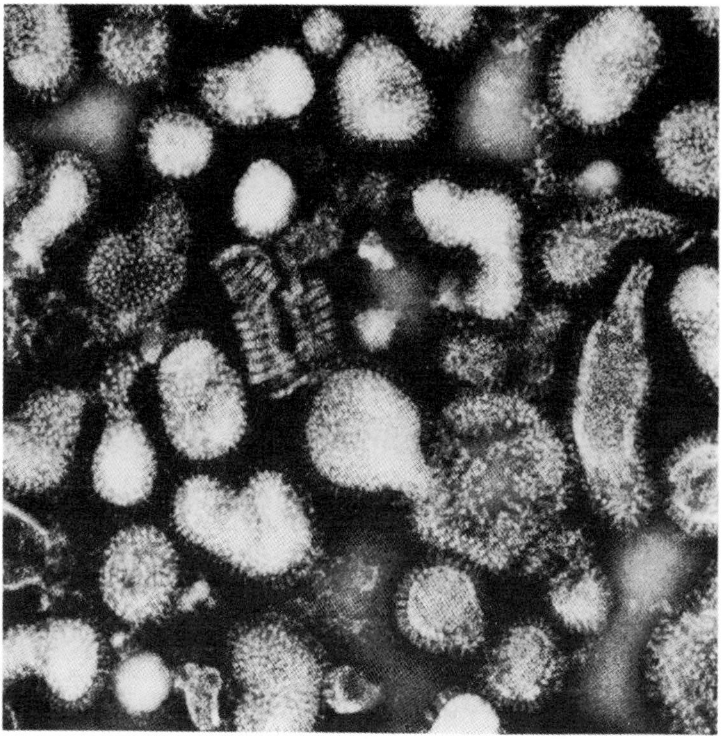

Virus particles of Asian 'flu, about 100,000 times larger than life, as seen in an electron microscope. The spikes visible on some of the particles serve as points of attachment for the antibodies of immune persons. (R. Dourmashkin.)

mulated a prediction: that any new outbreak of influenza would be very patchy. An epidemic in the winter of 1977–8 gave them the opportunity they needed, and the boarding schools had precise information on where pupils slept and on who contracted the 'flu when. The result was that the first book on medical astromicrobiology, Hoyle and Wickramasinghe's *Diseases from Space* (1979), was decorated with plans of schools and dormitories, like one of the more complicated detective novels.

And from the tribulations of the adolescent upper crust that winter they proved, to their own satisfaction at least, that 'flu was not communicated from person to person and was even more patchy than they predicted. For example at Eton College one third of the boys were taken ill while St George's College, situated just a few hundred metres away and sharing the same medical officer, had no influenza at all. The Eton boys slept in twenty-five different houses, but they mixed for teaching and to some extent for eating, yet the incidence of influenza ranged from a single case among seventy boys in one house, to levels well above the average in others.

The odds against these variations occurring by chance were rated by the astrophysicists as astronomical, and they offered a meteorological explanation for the fine-scale patchiness. The most effective viruses from outer space would be those washed down in raindrops that 'dried out' before they reached the ground, leaving the viruses to tumble down the rest of the way in local turbulence. The direction of the wind near the surface, interacting with topography and buildings, could greatly affect the patterns of turbulence, creating fortunate spots that escaped the virus.

It is unhealthy to go out of doors – that is an obvious deduction from evidence of this kind, and the theory in general. A tragic example is said to have been the Legionnaires' disease of 1976. Some members of the American Legion just stepped outside their hotel lobby in Philadelphia to watch a parade go by and were zapped by a small comet-cloud carrying a previously unrecognised bacillus. Twenty-nine failed to recover.

The particular suspicion that the comet *Halley* is responsible for influenza deserves a little elaboration. Hoyle and Wickramasinghe do not mean all variants of that disease but the two particular forms of the virus known to pathologists as H2/N2 and H3/N2: the first of these is Asian 'flu, which struck world-wide in 1957, while the second was responsible for the pandemic of 1968. In both cases there was evidence that many people over about seventy-five years of age had immunity to the virus subtypes, implying that they had been infected in previous outbreaks – a figure that jibes conveniently with the seventy-six-year period of *Halley* in its orbit around the Sun. (Also, you may think, with human longevity.) As explained earlier, the comet did not have to be anywhere near the Earth in 1959 or 1968: only its 'flu-laden clouds.

While you have a theory bubbling briskly on the front burner of the imagination, you may as well take care of the *Mary Celeste*, the Fall of the Roman Empire, and the Problem of Evil. Now as far as I know Hoyle and Wickramasinghe have not yet called down bugs from space to explain the disappearance of the entire crew of a well-found brig in mid-Atlantic, but they have no difficulty with the decline of Rome. During the period from AD 400 to 1400 the earthlings had a particularly nasty time with the clouds of diseases spun off from the comets. A 'disease-filled millennium', they call it, and its onset forced people to live farther apart and thus to 'uncivilise' themselves; it also , so they say, moved the Europeans to adopt the 'sombre' religion of Christianity.

As for the theological Problem of Evil, the apparently needless suffering from disease that God permits, it can be translated into a biological question. Why have we not evolved perfect immunity to infectious diseases? The answer given by the astrophysicists is that disease is necessary for major steps in evolution to occur, because organisms can incorporate novel hereditary messages from bacteria and viruses – not just mutant genes, but whole new sets of genetic material. Organisms that did not let the viruses in would stop evolving and soon become extinct. In this way the two authors make comets out to be a driving force throughout the history of life on Earth, and also attempt to disarm their medical critics who ask tetchily how viruses newly arriving from the depths of space know how to attack highly evolved animals like ourselves.

There was great rejoicing when the World Health Organization announced the elimination of smallpox from the Earth – the deadly and disfiguring disease that had scourged mankind since ancient times. But Hoyle and Wickramasinghe shake their heads gravely: they see an historical pattern in which smallpox comes and goes at intervals of several hundred years and declare that the evidence clearly points to smallpox coming from a cometary source.

Not even physiognomy escapes the astrophysicists' attention. The human nose, they say, has evolved its shape, with nostrils opening downwards, because of the protection it gives against viruses falling out of the sky. This sets me thinking: it used to be supposed that comet watchers and other stargazers were especially at risk from wells, road works and open manholes. Now it seems that those who turn their faces to the sky are also undoing the careful work of natural selection and courting all kinds of horrible infections. Brain fever, too?

No one in his right mind, other than manufacturers of protective suits, could wish this hypothesis of diseases from comets to be correct. Any gain in medical enlightenment would be more than offset by cometary hypochondria. But if, as well you may, you find the ideas of Hoyle and Wickramasinghe hard to swallow and perhaps entirely ludicrous, then attend closely to your reactions. It is an exercise in the archeology of emotions, like handling a shard of an ancient pot and empathising with the girl who was beaten for breaking it.

In this case we can feel at one with sceptics down the ages. How we should have enjoyed ourselves in Elizabethan inns, staggering about in our cups and blaming old Copper Knickers in Poland who'd put the whole world in a spin. Or in Victorian pubs: 'Does Mr Darwin claim descent from the monkeys on his mother's side or his father's?' These parallels, of course, say nothing for or against the theory of the 'flu from outer space, but only that scepticism is always good for a laugh.

If Hoyle and Wickramasinghe have made out a case for doctors to pay more attention to windborne viruses in the transmission of disease, even if the viruses are entirely earthbound, that is a serendipitous discovery indeed from studies of the chemistry of interstellar clouds and comets. This possibility reminds me of one of Hoyle's earlier efforts in his war on the Big Bang theory of the origin of the universe; he was at pains to prove, in association with sympathetic colleagues, that the chemical elements were made not in the primeval turmoil, but in exploding stars. He was right about that and present-day understanding of cosmic chemistry hinges on that classic study. The very atoms in my body and my pen teach me not to despise a theory just because it connects mundane events of today with cosmic cataclysms long ago.

The wish to 'prove Fred wrong', when he was champion of the Steady State theory of the universe, was also a powerful stimulus to the astronomers who, in the 1950s and 1960s, pushed their techniques to the limit in order to find out the overall nature of the cosmos. In disproving the Steady State, astronomy benefited greatly. Cometary and meteoritic research, too, is already being stimulated by the ideas of Hoyle and Wickramasinghe. Even scientists who are thoroughly sceptical about the origin of life in comets, never mind the diseases, are happy enough, when setting out research objectives for space probes, to add the unexpectedly grandiose one of searching for deep-frozen viruses and bacteria.

But anyone would hesitate to look for birds' nests in a blast furnace and naturally there is argument about the theory and its technical shortcomings. Those are the only shortcomings that matter, by the way: not tact about medical or theological sensibilities, nor even literary style. Hoyle and Wickramasinghe violate no laws of nature, but they do seem to stretch improbabilities of theory and evidence like bubble gum. Their identification of cellulose in interstellar dust now seems to be mistaken: there is not enough hydrogen in the dust. Or take those comet grains waltzing in space, and supposedly harbouring the mumps viruses that wait to ambush some child's jaw: if the grains are too big they will burn up as meteors in the atmosphere; if they are too small they will not protect the viruses against the sterilising ultraviolet rays from the Sun, which may de-mump them, while the pressure of sunlight will sweep them out of the Solar System. But for me credulity fails at the very start of this viral itinerary.

I really don't care for the supposition that out in the Öoo Cloud there are comets, which have never been near the Sun, that contain horrible diseases we have not thought of yet, perfectly adapted to infecting and prospering in the bodies of our remote descendants. The putative diseases from space have to be renewed because every scrap of junk – comets, meteorites, meteors, mumps – is cleaned away by gravity or impact in about a hundred million years. If the comet *Halley*, for instance, is the amazing interplanetary 'flu machine, space will be purged of its outpouring after about a million epidemics. So Hoyle's and Wickramasinghe's own rules of the game require a continual supply of new diseases from 'new' comets falling for the first time into the inner Solar System, with viruses targeted on humans, pigs and peanuts. Since this dislodgement happens to only a very small sample of comets from the Öoo Cloud, and one at least in each 'batch' has to be infected if we earthlings are to be kept sweating in the isolation wards, there must be many pathogenic comets which use the same genetic code as we do. Otherwise the viruses will be as ineffectual as a car key in a padlock.

To provide a universal genetic code, Hoyle and Wickramasinghe have to say that one comet scooped the pool. The first of them to evolve successful living things infected many other comets capable of supporting life. As a result they use the same basic chemistry, whether it has been their fate to seed the Earth with the first life or to skulk at the edge of the Solar System. They would have us believe that a billion comets out beyond Neptune contain viruses capable of invading humans and commandeering their cells to reproduce the cometary genes and clothe them in neat protein jackets, as worn this season in the Öoo Cloud. Cross-infection between comets is supposed to do the trick, but Hoyle and Wickramasinghe have already assured us that their nurseries of life are shielded from the outside by hundreds of metres of frozen crust.

Judicious astronomers and biologists in the early 1980s might rate the stages of the thesis as follows:

a. Comets supplied a rich soup of chemicals to the young Earth. *Plausible.*

b. Life can originate independently in the nucleus of a comet. *Doubtful.*

c. Our own microbial ancestors came to the Earth in a comet. *Doubtful squared.*

d. Some diseases are transmitted unexpectedly via the air, rather than from person to person. *Possible.*

e. Diseases sometimes help evolution along by introducing new genes. *Plausible.*

f. Diseases come from present comets. *Yes, and pigs can fly.*

These things are not settled by opinion polls, although the philosopher Imre Lakatos noted that prevailing beliefs in science are often a matter of mob psychology. Opinion might one day swing behind Hoyle, but that would not by itself make his ideas any the more true. Like democracy, science succeeds in the long run not because its luminaries are free of error, but because they are often found out; sooner or later nature puts an end to nonsense by delivering its verdict in a suitable experiment. A single bacillus of the Black Death, scooped up somewhere beyond the Moon, could change a lot of tunes.

Scientists are not invariably desperate to have their theories confirmed; some might sooner die not knowing the answer than risk being disproved. But as mortals they want acceptance and acclaim, which is a different thing. Unlike some professions, where you reach the top by living long enough and not annoying anyone important, in science you are rated by the amazing discoveries and inventions you are thought to have made. In the gamble of high research the best form is therefore seen in a theory sufficiently surprising and provocative to be memorable and yet not so eccentric as to be scorned throughout the author's lifetime.

Out of two broad categories of winning theories, one is the surface ploy, where the hypothesis is original and yet readily believable by a substantial group of colleagues; then it hardly matters whether or not the hypothesis is right or wrong. Newton's theory of gravity and the snowball theory of comets are examples of surface ploys. In a deep ploy, on the other hand, the hypothesis is obviously incredible and everything is then staked, at long odds, on showing that it is correct. The theory of continental drift, like the hypothesis of germs from comets, was a deep ploy. It is a much more amusing *coup* if you can pull if off, but Alfred Wegener died before he was vindicated, and that is why people like Sir Fred Hoyle make intense and urgent gamblers: posthumous acclaim buys no beer.

A thoughtful observer of the scientific betting shop, the biologist Sir Peter Medawar, has said: 'I cannot give any scientist of any age better advice than this: the intensity of the conviction that a hypothesis is true has no bearing on whether it is true or not.' But as Medawar goes on to note, conviction is an incentive to work. Science is one of the most passionate of human activities: how else would researchers be sustained through the long weeks or years of drudgery, why otherwise should Hoyle and Wickramasinghe spend so much time in correspondence with school matrons? If appearances contradict this, it is because all gamblers pride themselves on keeping their outward cool.

Apropos mob psychology: another idea in astronomy, every bit as far-fetched as Hoyle's contaminating comets, is entirely fashionable. Non-scientists might suppose that 'orthodox' opinions in science are confined to propositions that are securely founded in facts (like the fossil record of evolution) or powerful theory (like the black holes of General Relativity) or at least in prehistoric habits of thought (like the geology that denied continental drift). This is not the case with a notion that is quasi-religious dogma, especially among American and Soviet astronomers.

Its world-wide influence was noticeable in England in 1968 when the Cambridge radio astronomers detected regular pulses coming from among the stars. They wondered if they should tell the Prime Minister and they labelled their record 'LGM' for Little Green Men. In fact they had discovered the famous pulsars, the pulsating stars, but their suspicion was aroused because of the newly fashionable belief that there are a million advanced civilisations in the Milky Way, alien folk who are supposed to be busy conversing with one another by radio and looking out for primitive newcomers like ourselves.

Many of the most respected names in modern astronomy subscribe to this idea. The US National Aeronautics and Space Administration and the Soviet Academy of Sciences run projects in support of it, and spacecraft now leaving the Solar System carry messages in bottles. You may debate, if you want to, whether the number of races of communicative superbeings is a thousand or a billion. You are allowed to haggle about the best strategy for detecting their signals and how to wheedle money out of governments for the search. You can even question an assumption of galactic anthropology that they are all good guys who mean us no harm. But the way to make yourself thoroughly unpopular is to suggest that no one at all is trying to get in touch with us.

Arguments against the existence of the Galactic Club are brushed aside. One is the biological view that the peculiar sequence of events which produced chatterboxes with radio telescopes on the Earth is less likely to be mimicked elsewhere in the universe than the astronomers imagine. Another is the engineering argument that an advanced civilisation would colonise the entire Galaxy by starship within a few million years, as we ourselves can expect to do. So where are they? Unless you have poor enough eyesight to see flying saucers, the aliens are conspicuous by their absence from the Solar System. Astronomers ridicule the UFO buffs but share their basic belief in the messiah from outer space, the little chap with goggle eyes far wiser and cleverer than ourselves, who will teach us the ways of righteousness and peace. The only difference is that saucerers claim to have met him already.

Whether there are more alien intelligences in the Milky Way than plagues among the comets, only time will show. Meanwhile my whiskers twitch as I study the comments of a leading American astronomer, well known to the public for his colourful language and bold hypotheses, about a paper written by a young scientist who argues that their existence is unlikely. Every proposition, however reasonable, is condemned and the whole thing smacks of a priest nosing out heresy. These comments come from the hidden world of refereeing, wherein scientists decide if one another's papers are worth publishing. In that connection two psychology teachers at Grand Forks recently took some papers already published in leading journals of psychology, retyped them, altered the authors' names and addresses, disguised the first few sentences, and resubmitted them to the very journals in which they had already appeared. In most cases the plagiarism passed unnoticed and three quarters of the reviewers and editors declared the articles unfit for publication.

The pantomime never ends. Some years ago I experienced the heavy-handed scorn and vilification with which entrenched scientists greet unwelcome discoveries. It concerned the rhythm and causes of past ice ages, and the imminence of the next one; I had contributed a scrap of original research but mainly given publicity to magnificent work by others. The ensuing row cost me a year of broken sleep, until what we had reported was well confirmed and it became the accepted story. We might have been wrong and such minor ordeals now seem to me unavoidable. Scientists in pursuit of enlightenment from nature will always behave like men competing for the same woman.

To all honest scientists I commend Alfred Tennyson's reply to a spiteful critic:

> I think not much of yours or of mine:
> I hear the roll of the ages.

And as it is hardly appropriate for bystanders in the pantomime to heckle, I try to moderate my ridicule of the theory of diseases from comets. Yet I suspect that comets have shaped the course of life on this planet less by injecting new genes than by killing off old ones.

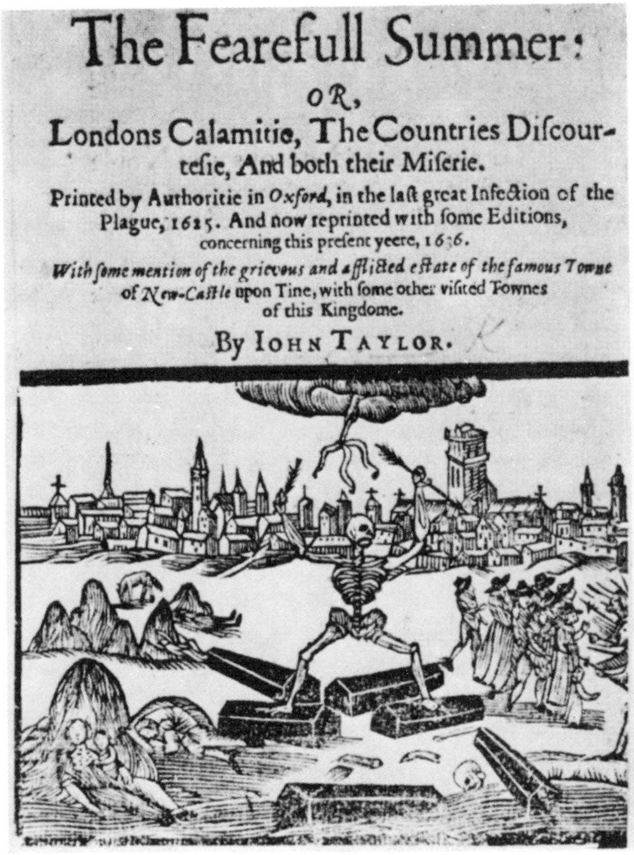

The Fearefull Summer:
OR,
Londons Calamitie, The Countries Discour-
tesie, And both their Miserie.
Printed by Authoritie in Oxford, in the last great Infection of the
Plague, 1625. And now reprinted with some Editions,
concerning this present yeere, 1636.
With some mention of the grievous and afflicted estate of the famous Towne
of New-Castle upon Tine, with some other visited Townes
of this Kingdome.
By Iohn Taylor.

In seventeenth-century England there was good reason to fear the Plague. If diseases came from space, such visitations would depend on fluctuating deliveries of pathogens from comets, and so could easily recur.

7. COFFIN OF THE DINOSAURS

Edmond Halley thought that Noah's Flood was caused by a comet. The train of thought that carried him to this unlikely terminus started off in 1677, when the young Halley was mapping the stars of the southern sky during an expedition to the island of St Helena in the tropical Atlantic. There he found the heavy night-time dew soaking his paper 'that it would not bear ink'. His pioneering work in southern-hemisphere astronomy was impeded, but his embarrassments with the soggy paper of St Helena inspired him to make researches into dew and atmospheric water in general. Back at home in England he measured the evaporation of water from pans, computed the flow of the River Thames, and generally busied himself with questions about the water cycle.

Armed with new information in hydrology and having, in addition, some grasp of the global climate, Halley then turned to prehistory. In particular he addressed the meteorological problems raised by the account of Noah's Flood in the book of *Genesis*. Like many men of learning of his time he looked upon the fossil seashells found in hilltops as evidence for a very great flood in the remote past:

> That some such thing has happened may be guessed, for that the Earth seems as if it were made out of the ruins of an old world, wherein appear such animal bodies as were before the Deluge.

So although Halley made bantering remarks about 'the agreement of the animals among themselves' in Noah's Ark, he took the supposed geological event quite seriously.

Troubled by the hydrological impossibility of flooding the world and covering the mountains with a mere forty days' rain, however heavy, Halley was moved to suggest to the Royal Society in 1694 that the Deluge was due to 'the casual shock of a comet'. A comet strike might alter the Earth's axis of rotation and a mighty sloshing of the oceans could ensue, which would scour the sea bed:

A nineteenth-century French caricature of a recurrent human fear: a comet hits the Earth and smashes it. (American Museum of Natural History.)

. . . for such a shock impelling the solid parts would occasion the waters . . . with one impetus to run violently towards that part of the globe where the blow was received; and that with force sufficient to rake with it the whole bottom of the ocean, and to carry it upon the land; heaping up into mountains those earthy parts it had borne away with it . . .

Then, Halley suggested, huge waves would recoil, 'reciprocating many times'. He offered the Caspian Sea as a possible comet crater, long mountain ranges as places where opposing waves converged, and north-east Canada as a former North Pole, where the ground was still frozen.

Hindsight of the astigmatic kind has led some biographers of Halley to pass hastily over this feverish outburst from their hero. But by drawing attention to possible consequences of a collision between a comet and the Earth, Halley triggered an unending succession of pseudoscientific theories and prophecies of disaster that put publishers for ever in his debt. And such is the perversity of nature, this seeming whimsy is of greater scientific significance than his prediction of the return of a particular piece of orbiting trash.

An unseen piece of a comet hit the Earth seventy-odd years ago and killed some wildlife; it was very much smaller than *Halley* and many other comets. The bombardment of the inner planets recorded on the face of the Moon has not entirely ceased and, at very long intervals, large comets or their residues can gravely injure the Earth. Saying so tends to create panic among people with no sense of the chasms of geological time that have opened up since the Bible was written, and little appreciation of the vast scale of the Solar System. If Satan is hurling snowballs at us from the Öoo Cloud, his aim is unimpressive.

Halley postponed publication of his Royal Society paper on the Deluge for thirty years, fearing *ultra crepidam* (a cobbler should stick to his last) and also 'censure of the Sacred Order'. But some in the Sacred Order were keen that, in the aftermath of the copernican revolution, the Bible should be validated in cometary terms. Eighteen months after Halley's talk, the cleric and mathematician William Whiston of Cambridge modified and embellished Halley's version of events.

In his bestseller *A New Theory of the Earth*, Whiston related how the Earth itself began its existence as a comet that evolved into a planet. A second comet with 'little or no atmosphere' struck

A twentieth-century concept of disaster from the sky; New York is a prime target for fictional events of this kind.

the Equator and set the Earth spinning. Then human sinfulness brought a third and punitive comet which passed close in front of the orbiting Earth at noon, Peking time, on Monday 2 December 2926 BC. The tidal force cracked the Earth and the tail of the comet ('a sort of mist') drenched it with water six miles deep:

> . . . a comet is capable of passing so close by the body of the Earth as to involve it in its atmosphere and tail for some time and leave prodigious quantities of the same condensed and expanded vapours upon its surface.

The comet in question was the very one that Whiston and his readers had seen in 1680–1, and Halley himself had suggested was on a 575-year orbit. And as for 'the ancient drowning', so for 'the future burning': a comet would set the world on fire. If it were to be the same 1680 comet, that would be in the twenty-third century AD. In all this Whiston sought to prove the veracity of *Genesis* and *Revelation*. Voltaire reported at the time from England, and told his French readers that Whiston was 'unreasonable enough to be astonished that people laughed at him'. Not everyone did, and some mistook the planet Mars, on a close approach in 1719, for a comet on its way to finish them off.

Whiston's speculation was not entirely stupid, at a time when the Bible was the main source of information on prehistory. Similar scenarios were frequently advanced by later authors, but by the mid-twentieth century it was clear that *Genesis* need not be taken too literally. Radioactive dating showed the Earth to be some billions of years old, while fossil animals told of the slow, majestic evolution of life. But there was still money to be made out of human credulity. The silliest comet theory of them all reads like a parody of Halley and Whiston; loyal Velikovskians would say it even anticipated the current propositions about life in comets.

Not just viruses but fully-fledged flies and frogs were brought to the Earth by a comet, according to the psychoanalyst Immanuel Velikovsky, who died in 1979 lamented by a large band of devotees. He explained many miracles of the Old Testament by invoking an immense comet that was hurled from the planet Jupiter. It brushed past the Earth more than once, producing at will not only Noah's Flood but also the parting of the Red Sea and (in a neat inversion of Whiston) the Joshuan trick of making the Sun stand still, by stopping the Earth's rotation. After rampaging about for a while, scattering animals to plague Egypt and carbohydrate manna to feed Israel, the comet settled down in a circular orbit, so Velikovsky said, and became the planet Venus.

Unlike Hoyle and Wickramasinghe, or even Whiston in his day, Velikovsky was indifferent to well-established facts and theories. Non-scientists complain that the astronomers responded to Velikovsky's *Worlds in Collision* with excessive fury and dogmatism. Of course they did but astronomers, as we have seen often enough, are only human. How would historians react, I wonder, if a psychoanalyst grew rich on the proceeds of a book asserting – as 'fact', not fun – that Napoleon was in reality the woman depicted in Leonardo's Mona Lisa who, after an affair with Karl Marx, gave birth to George Washington?

If cowards die a thousand deaths so do the credulous, and definite expectations of 'the casual shock of a comet' prevailed in Paris in 1773 and in London in 1857. While the French were thrown into panic by misbegotten rumours that a comet hunter was expecting an impact, the British scare was more deliberate. A sixpenny book went on sale in London, with the title *Will the Great Comet now Rapidly Approaching Strike the Earth?* It predicted the reappearance of a comet seen in 1264 and 1556, and warned the public to beware of the great heat that would result. In 1858 the comet *Donati* duly appeared, but from a direction completely different from the earlier comets, which were in any case not the same. It missed.

In fiction not pretending to be fact, writers subject our poor planet to a rain of terror. The comet *Hamner-Brown* destroyed civilisation by direct impacts, huge tidal waves and a nuclear war which it sparked off between Russia and China, in *Lucifer's Hammer* (1977) by Larry Niven and Jerry Pournelle. In the misnamed movie *Meteor* (1979) a comet smashed the main-belt asteroid Orpheus and a fragment neatly scythed the skyscrapers of New York City, although the world was saved from the main impact by concerted use of American and Soviet nuclear weapons. The most convincing feature of that story was the large amount of liquor consumed by the astronomers monitoring the approach.

The effects of fictional comets were not always entirely unfavourable: in *Olga Romanoff* (1894) George Griffith neatly disposed of his invincible female villain by a cometary bolt from the blue, while *In the Days of the Comet* (1906) by H. G. Wells had a benign vapour from a comet's tail causing an outbreak of reason, peace and love throughout the world, in a pre-Velikovskian utopia. But the real collision with a small comet in the first decade of this century went unnoticed. Later it became infamous enough

In 1908 a small piece of a comet hit Siberia and the blast wave flattened the forest over thousands of square kilometres.

and the Tunguska Event reverberated for decades in the imaginations of the scientific and pseudo-scientific worlds.

Early in the morning of 30 June 1908 the driver of the Trans-Siberian express heard loud bangs and imagined that his train had exploded. When he stopped, his wide-eyed passengers said they had seen a bright blue ball of fire streaking across the sky, trailing smoke. Six hundred kilometres away to the north-east, in the valley of the Podkamennaya Tunguska river, a blast uprooted huge areas of forest. It slaughtered reindeer and scattered the tents of nomads camping far from the explosion. In present-day terms, it was like an H-bomb going off. Experts hearing the news suspected a big meteorite but inconvenient wars and revolutions prevented them reaching the scene until 1927. Then, and in subsequent Soviet expeditions, they found the shattered forest but no large crater, only a number of small holes and some meteoritic grains a tenth of a millimetre in diameter.

The strange goings-on in Siberia were therefore open to any outlandish or otherworldly explanation. When the physicists discovered anti-matter people suggested that a chunk of that deadly stuff, annihilating ordinary matter, had caused the Tunguska Event. A distinguished British nuclear-weapons maker supported the suggestion that a natural nuclear bomb fell out of the sky at Tunguska. Flying saucers became popular, so the Siberian backwoods were flattened by an alien spaceship crashing, or taking off. No sooner had astronomers become interested in black holes than one of those was said to have bored through the Earth: in at Tunguska and out through the Atlantic.

By far the most plausible explanation of the Tunguska Event was that a small comet hit the Earth. This suggestion originated in 1930 with Francis Whipple of Kew, London, not to be confused with Fred Whipple, Snowball Maker, of Cambridge, Massachusetts. But the rise of the snowball theory of comets encouraged that view of Tunguska, and by the 1960s Soviet scientists were inclined to agree with it. In 1975 an Israeli scientist, Ari Ben-Menahem of Rehovot, reassessed all the information and concluded that the main explosion occurred 8.5 kilometres above the ground and was equivalent to 12.5 megatons (million tons) of high explosive going off. That equals a moderately large H-bomb. To cause such a blast, David Hughes of Sheffield calculated that the impact on the atmosphere of a Whipple-Whipple snowball a mere forty metres in diameter, and weighing about 50,000 tons, would be sufficient.

The absence of large stones and craters makes sense if the comet consisted mainly of ices. That it was not spotted in space before it hit is unsurprising: so small a comet would not be visible to the naked eye until a few minutes before impact. Material shed from

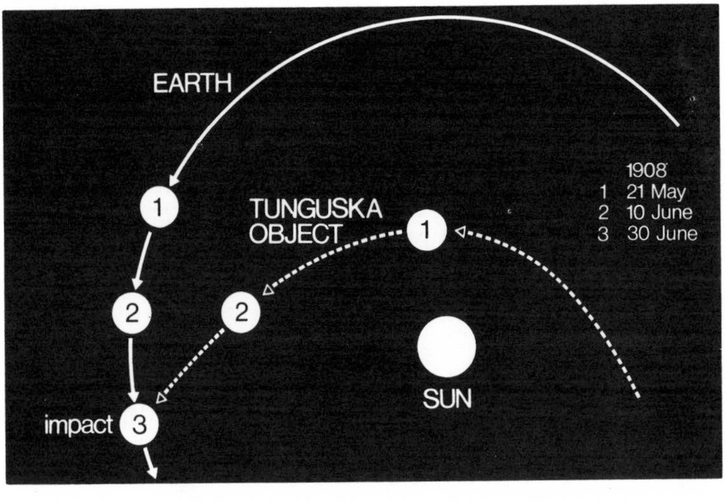

The track of the micro-comet, judged to be a piece of Encke's Comet, as it hurtled towards a collision with the Earth, at Tunguska in 1908. (After L. Kresak.)

the comet as it sloped in through the atmosphere above Europe and Asia explains a mysterious brightening of the night sky noted in those regions in July 1908. There was just one snag. An American Nobel prizewinner had supported the proposition that the Tunguska Event was caused by a body containing anti-matter, by saying that the amount of radioactive carbon in the Earth's atmosphere was increased by the event. Hughes and a colleague went to some pains to account for the radiocarbon, until a letter from the famous man said he meant the opposite: there was no increase in radiocarbon. Another candidate for the sperm bank, it seems.

The Tunguska comet coincided with a daylight meteor shower consisting of dust particles left in the orbit of the comet *Encke*, so it was probably a very small fragment of that comet. Lubor Kresak of Bratislava has made out the detailed case for this identification. If the whole nucleus of *Encke*, 100,000 times more massive, hit Siberia, it would kill more than reindeer. But the threat to the Earth comes not just from the active comets that brandish their heads and tails around the Solar System, but from the small, dark apollo objects, the micro-planets that cross the Earth's path.

On grounds of mental hygiene there might be a case for playing down the risks of cosmic collision. Comet lovers sometimes attempt a cover-up, seeking to persuade their fellow citizens that comets are merely pretty. To see them as obnoxious objects which ought to be brought under control accords better with the facts.

A roulette player is sure to win now and again if he goes on gambling for ever and impacts from the snowball warheads of comets are certain to occur during the longwinded timescales of geology, although the odds against any one comet on a random orbit hitting the Earth are something like a billion to one against. Even a comet on a tight orbit that brings it repeatedly across the Earth's path is not very likely to hit us while it still glows. But live comets may have left their fingerprints on the surface of the Moon, in the form of whorls of brightness that straddle the old slopes and craters like party streamers dropping where they will. Scientists in Houston relate them to streamers seen in the heads of comets – zones of dust and gas that struck the Moon at high speed. Comets are strong candidates for the most violent collisions with the Earth, but apollos are much more likely to score a hit.

The orbits of the apollos are strikingly similar to that of *Encke* and when that comet has exhausted its supply of volatile materials, after too many visits to the Sun, it can be expected to become a dark micro-planet comprising a stony and tarry residue, perhaps a kilometre in diameter: an apollo object in fact. Ernst Öpik has for long argued that apollos are dead comets. Now George Wetherill of Washington reckons the supply of comets and pronounces it more than sufficient to account for all of the apollos – whether visible or conjectured. Gravitational football and jet propulsion have to put only one comet into an *Encke*-like orbit every 65,000 years or so.

Astronomers have spotted about thirty apollos so far, crossing and recrossing the Earth's orbit in a free-ranging fashion, and they infer the existence of many hundreds of such objects, a kilometre or more in diameter, circulating in the inner Solar System. They are hard to see because they are small and dark and in 1937 Hermes came out of nowhere and scampered past us at only twice the distance of the Moon. These comet corpses must count as the most unpleasant form of sky pollution, because more than one in four of all the apollos is due to collide with the Earth sooner or later.

As Wetherill judges the celestial roulette to proceed, *Encke*-class comets turn into apollos at a rate of fifteen every million years. Of that fifteen, gravitational football ejects seven, harmlessly. Half of the remaining eight collide with Mercury, Venus, Mars or the Moon. Four collide with the Earth. In other words, once in 250,000 years, on average, a micro-planet hits the Earth. Typically the object is one kilometre in diameter and it excavates a crater about twenty kilometres in diameter. The energy of impact is equivalent to the explosive force of 100,000 megatons (million tons) of TNT. Much less frequently, perhaps once in a hundred million years or so, the impacting object is ten kilometres across and the explosion is a thousand times greater.

In space collisions the energy of motion is converted into explosive energy, and each ton of comet ice or apollo dirt becomes equivalent to far more than its own weight of high explosive – anything from a minimum of fourteen tons of TNT when the alien object is barely captured by the Earth's gravity, up to more than a hundred times as much if the object is a long-range comet travelling at high speed across the Earth's orbit. The morbid formula for the energy is $\frac{1}{2}mv^2$, where m is the mass and v^2 is the square of the velocity of impact.

For some decades imaginative experts have suspected that the shocks of successive comets and apollos were responsible for upheavals in life on Earth discernible among the fossils. It now looks as if this supposition is right, and the chroniclers of life, who have so recently had to adjust their ideas to the confirmation of one crazy idea, continental drift, are now faced with another. As with ice ages and black holes, the discoveries of factual science turn out to be more dramatic, and the mental pictures they evoke more horrible, than any conjured up by fiction and pseudo-science.

Admirers of Edmond Halley may be grateful that he did not pursue a career in botany. His only proposal in that direction was an experiment for seeing if garden plants would prosper when you blotted out the light of day completely, using sheets of brown paper on glass, which is like wanting to know whether a bird will sing any differently with its head cut off. It smacks of the projects seen by Jonathan Swift during a visit to the Royal Society in 1710, when Halley was a Fellow of thirty years' standing and soon to become the Society's Secretary. In *A Voyage to Laputa* Swift describes 'projectors' at the Academy of Lagado, whom Gulliver finds busy with designs to sow land with chaff, propagate a breed of naked sheep, generate books in philosophy by random combinations of words, and extract sunbeams from cucumbers.

With only moderate exaggeration Swift captured the enthusiastic ignorance of the eighteenth century. Halley's brown-paper project predated all sound experimental knowledge of how plants use the radiant energy of sunlight to combine carbon dioxide and water, and so make the food upon which, directly or indirectly, all animal life depends. That process has continued without interruption since the early days of life on Earth. Or has it? Three centuries after Halley it turns out that a comet, alive or dead, is capable in effect of wrapping our entire planet in brown paper – a dismal feat which kills most living things. And that, it seems, is what befell the dinosaurs.

Facing: Light-coloured patches and swirls strewn among the craters may record a comet hitting the Moon. The larger of the two small craters in the middle of the picture is thought to mark the impact of the comet nucleus. (P. Schultz and L. Srnka, NASA.)

Comets and dinosaurs have in common an extravagant size, long tails and a titillating hint of terror. If people go crazy about comets they are not a little demented about dinosaurs, and to say that a comet killed the dinosaurs threatens to compound the manias. But when grim manuscripts arrived from Berkeley and Amsterdam, I was compelled to fly off and see the coffin of the dinosaurs for myself. It once encased the world and shrouded the bones of the last of the dinosaur dynasties. In the sixty-five million years since the event, geological processes have buried,

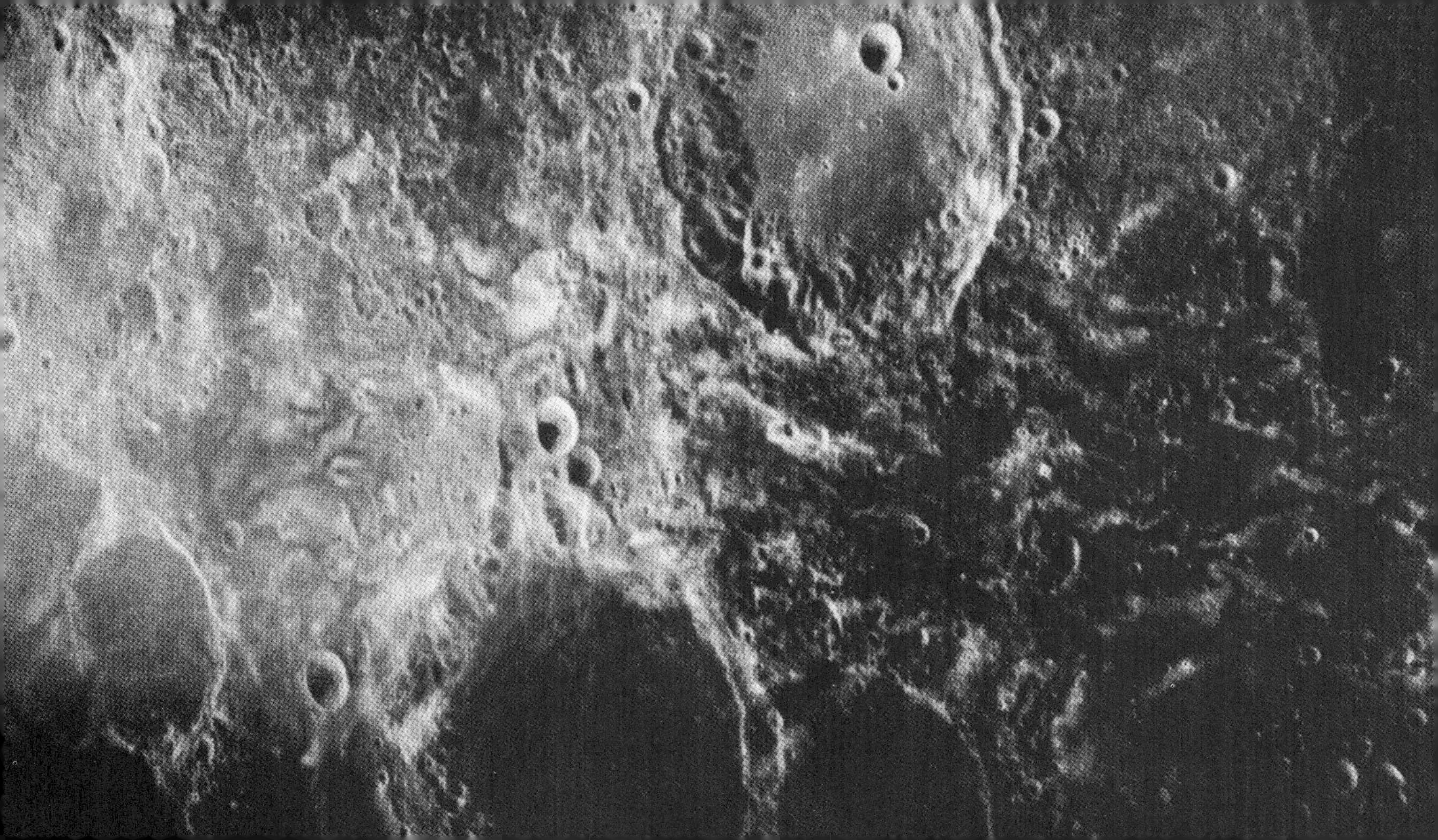

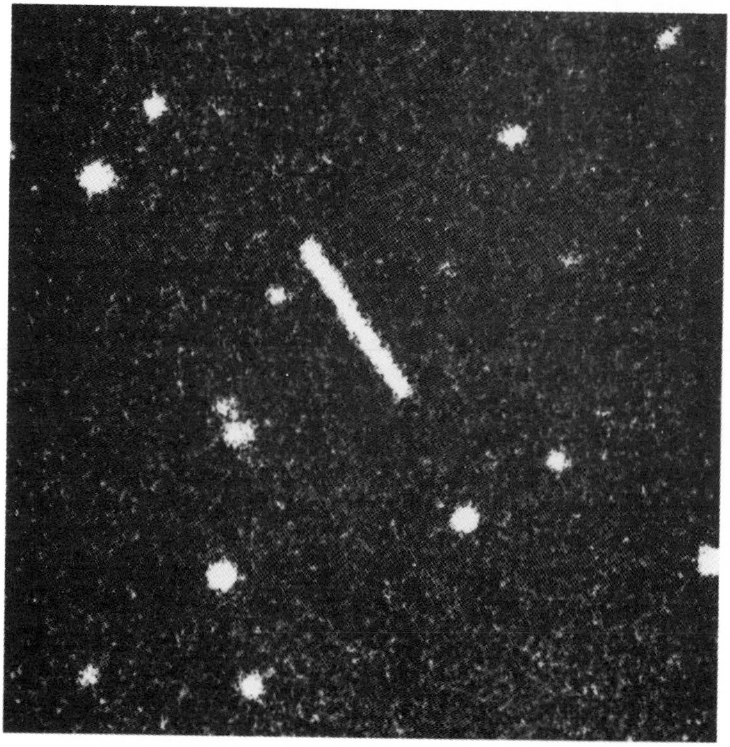

Apollo objects – small, dark-coloured bodies crossing the orbit of the Earth – are hard to see, but their movement relative to the stars can give them away by producing a streak on a telescope photograph. This one is Ra-Shalom, so named because it was detected in 1978 at the time of Middle Eastern peace talks. (Eleanor Helin; Palomar Observatory photograph.)

eroded or mangled it. It is best preserved in an accessible form where Earth movements have pushed up into the air the well preserved sediments from the bed of a former sea. One such place is Umbria in Italy, so I found myself there, in the small medieval city of Gubbio.

A geologist from Perugia, Giampaolo Pialli, took me to the gorge just outside Gubbio, where the river conveniently carved through the rocks and exposed the layers of ancient pink limestone. They had been tilted at a giddy angle, like a stack of playing cards towering beside the road, and they recorded many millions of years of Earth history. Up the gorge we went, through the layers of the geological periods called the Jurassic and Cretaceous. During all this time and more, the reptiles ruled the disintegrating supercontinent of Pangaea.

My guide eradicated all vague conjecture as he pointed out the many places where rock samples had been taken, for evidence of the repeated reversals in the direction of the Earth's magnetism. Correlated from place to place around the world, the patterns of flipping magnetism helped to pinpoint corresponding events with great precision. Incidentally, they indicated that the death of the dinosaurs did not coincide, as some would have it, with a geomagnetic reversal. The late Cretaceous rocks bore no hint of the fate in store for the dinosaurs. And then there it was, sloping up the wall of rock, neat as a line across a geologist's time-scale: the coffin of the dinosaurs.

It was a layer of red-brown clay, barely a centimetre thick. Beyond it the limestone layers resumed, but they were of the Tertiary period, when the dinosaurs were extinct. By the roadside, the sample-points of the palaeomagnetists bracketed the coffin-layer like bullet holes. When I reached and picked out a flake of clay, I knew that it was rich in material not of this Earth. Looking from the clay in my hand to the cloud-harassed Sun, I tried to visualise the clay as fine powder hurled into the air by a colliding comet or apollo, and turning day to night.

The Case of the Disappearing Dinosaurs is of interest far beyond the little realms of comet lovers and dinosaur hunters. At the end of the Cretaceous period the giant animals that had dominated the scenery for more than a hundred million years, munching leaves or chewing meat, suddenly were gone. Had that event not occurred, we the mammals would still be shrew-like beasts cowering in crevices and branches to avoid those cruel teeth and claws; our beady little eyes might not even have noticed the comets.

As soon as Victorian fossil-hunters appreciated the downfall of the dinosaurs, explanations proliferated. The most concise theory was that the dinosaurs were too large to fit into Noah's Ark. The most persistent said that they were too stupid to survive and British campaigners against nuclear weapons *circa* 1960 employed the slogan: 'Dinosaurs died out: too much armour, too little brains.' The libel lasted into the 1980s, in a notorious poster in which a German car manufacturer likened his rivals' products to decrepit dinosaurs destined for extinction; it provoked reptilian hisses from dinosaur lovers and the advertisement was withdrawn.

Hollywood producers had women cringing in deerskins while their menfolk battled with the great brutes in the stone age, thus making good the sixty-million-year lacuna between the death of the dinosaurs and the rise of woman. And those who believed that God was an astronaut had no difficulty in imagining an alien Buffalo Bill massacring the herds of sauropods with laser beams, tossing his californium hand-grenades into the lairs of the tyran-

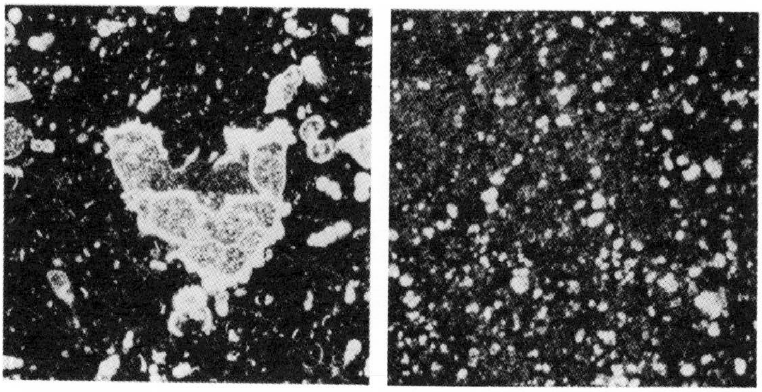

The disaster that struck small marine animals at the same time as the dinosaurs perished is apparent in these microscope photographs of fossils in the rocks of the Bottaccione Gorge at Gubbio, Italy. The first picture shows some foraminifers, about half a millimetre wide, living at the end of the Cretaceous; after the event a new, much smaller species took their place. (W. Alvarez, University of California, Berkeley.)

nosaurs, and loosing his surface-to-air missiles at the pterodactyls, which lost command of the air at the same time. But you would have to allow that this interloper also had a gargantuan appetite for shellfish.

A sea-change overtook the planet at the end of the Cretaceous period. The ammonites, well known to fossil collectors as coil-shelled jet-propelled creatures, perished just as decisively as the dinosaurs; so did other groups of small marine animals, and notable sea-monsters like the plesiosaurs. Monomania about the dinosaurs made people forget the ammonites, but the simultaneous termination of a wide variety of life on land and sea devalues any solutions peculiar to the dinosaurs. Thus even if the newly-evolved caterpillars destroyed vegetation as thoroughly as locusts, or rat-like mammals developed a hearty appetite for dinosaurs' eggs, as a couple of theories would have it, neither would affect life beyond the beaches. The same objection applies, unfortunately, to the most imaginative of all explanations of the dinosaurs' demise, which related it to the rise of the modern flowering plants. The involuntary change in diet, it was said, deprived the dinosaurs of the laxative oils that are present in the older conifers, ferns and cycads, so they all perished of constipation – no small matter in a dinosaur.

Desperate problem-solvers turned to physical causes: a drastic fall in sea-level, a global cooling, atomic radiation from space breaking in when the Earth swapped its magnetism around, the explosion of a nearby star. A phenomenon crying out for definitive explanation was just as creative of theories as any comet. Ten years ago the same clutter of a hundred contradictory hypotheses had surrounded the ice ages. For them, the correct explanation came largely from analyses of the atomic composition of seabed fossils, which revealed the real rhythm of the ice ages. So I was prepared to believe what the manuscripts had told me: that the atomic composition of the red-brown clay from an ancient seabed, which I held in my hand in the gorge at Gubbio, gave the definitive answer to the dinosaur problem.

Microfossils studied there and at similar sites by a generation of scientists put the time-scale of the crisis between ever-narrower limits. At Caravaca in south-eastern Spain, where the pattern in the rocks is very like Gubbio, Jan Smit of Amsterdam recently established that the teeming marine life stopped suddenly. The plankton, constituting the pastures of the sea, disappears in less than five millimetres of sediment – meaning about a hundred years, at the longest. Thereafter the surface waters remained a desert for ten thousand years. Before there was any appeal to atomic analysis the fossils themselves announced that the event that ended the Cretaceous period was quick and catastrophic, and dismissed all hypotheses proposing gradual changes in climate, ecology or bowel-movements.

At the University of California at Berkeley, an eminent physicist, Luis Alvarez, joined forces with his son, Walter Alvarez, a geologist who was familiar with the crucial layers at Gubbio. Their initial idea was to use delicate methods of chemical analysis as another way of finding out, not what killed the dinosaurs, but how long the crisis lasted. They reasoned that meteoritic dust, raining invisibly but continuously on to land and sea, was comparatively rich in iridium, a metal that is very scarce in the ordinary materials of the Earth's crust. The amount of iridium in the thin layer of clay at the end of the Cretaceous might therefore indicate how long the clay took to form. Samples from Gubbio went into a research reactor at Berkeley, to make them radioactive, so that iridium and other elements could be fingerprinted by the gamma rays they then gave off.

The result came as a shock: the amount of iridium jumped by a factor of thirty in the Gubbio clay, compared with adjacent limestone. Either the clay layer took an improbably long time to form, or the Earth was suddenly swamped with iridium. Even more startling results came from the same clayey layer at other places. Danish samples brought by Walter Alvarez to Berkeley showed the amount of iridium increasing 160-fold. Then the Dutchman, Jan Smit, collaborated with a Belgian atomic and meteoritic expert, Jan Hertogen, and they found, in the bottom part of the Caravaca boundary layer, iridium levels 460 times normal. The discovery was as suspicious as finding traces of arsenic in a dying man, the patient in this case being the sick planet Earth.

A forensic team consisting of Alvarez, Alvarez and two Berkeley space scientists, Frank Asaro and Helen Michel, soon satisfied themselves that the iridium could not have come from any plausible source on the Earth itself. They considered whether the popular idea of an exploding star might explain a sudden influx of extraterrestrial material, but calculated that the star would need to be improbably close to account for so much iridium. In any case, the other symptoms of a stellar explosion were lacking, plutonium atoms for example. So they were led to suggest that an apollo object struck the Earth sixty-five million years ago.

The Berkeley scientists' sums led them repeatedly, by different routes, to an apollo about ten kilometres in diameter, just a little bigger than the largest known apollos. Hitting the dinosaurs' planet, it would have thrown up about a hundred times its own weight of material from the Earth's crust. A fraction of the debris, scattered as dust all around the world, made the layer of clay, doped to just the right degree with the tell-tale iridium from the apollo.

In accordance with the Krakatau Effect, the dust would take a long time to settle. When that volcanic island blew up in 1883, dust high in the atmosphere produced 'glorious' sunsets all around the world for more than two years. A large apollo hitting the Earth causes an explosion more than a thousand times greater than Krakatau. A mushroom cloud sweeps material into the stratosphere, in the form of very fine grains that take years to fall out. The quantity of dust is more than sufficient to blot out the Sun completely and for about four years there is unending night.

Plants stop growing and the ensuing famine explains the disappearances among the fossil species. In the sea the microscopic plants die out almost completely and, if this is the true picture of the natural disaster at the end of the Cretaceous, the death of the oceans consigns to oblivion the ammonites, the plesiosaurs and other conspicuous marine animals; the survivors are presumably those that can scavenge in the mud for the remains of plants of former years. The putrefying remains of plants and animals, floating down the rivers, sustain some freshwater animal life and, as the sole survivors of the great reptiles, well may the crocodiles shed tears.

The plants on land are better equipped to recover afterwards, but during the fatal period all new growth ceases. Blundering about in pitch darkness the plant-eating dinosaurs strip every last leaf. The meat-eaters make a regal banquet of their herbivorous cousins and then, enraged by hunger, they turn cannibal. A few small animals – worms, insects, birds and so on – survive the four-year night feeding on decaying vegetation, seeds, nuts and one another. Among them are some of the diminutive mammals, which will found new dynasties in the depopulated world, when the sunlight breaks through again.

'Absolute nonsense,' a friend of mine scoffed. As an authority on the great reptiles, he had read only the early press reports on the supposed impact. His immediate objection was that some quite large reptiles clung on, and survived for millions of years after the end of the Cretaceous period. 'What about the pelomedusid turtles? Or the dyrosaurid crocodiles, six metres long? I've dug them up with my own hands.' His eyes glowed like cinders, as if his manual labours settled the argument.

Quietly I told him how I had sampled the layer of comet clay at Gubbio with a finger. I rehearsed the evidence for a very large and unusual event and spoke of the same clayey layer showing up in Italy, Spain, Denmark, Tunisia and France. 'Oh, of course,' he conceded, 'it's obvious that something happened, but it can't be as simple as they say. We'll need palaeontological control.' In less than five minutes my friend had switched from total disbelief to staking his specialism's claim to a piece of the action. He was quite right, of course: fossil hunters will have to trace the complicated consequences of the impact and make sense of the survival or disappearance of each kind of plant and animal. And to pretend that the story of life on Earth is just a succession of adaptations to catastrophic comet-shocks would be at least as foolish as to ignore the inevitability of intermittent events of that kind.

In 1980, when the Berkeley discoverers published their hypothesis, they stressed that it was unproved. Jan Smit in Amsterdam called an impact 'the most attractive' explanation. But the first severe test for the idea was soon fulfilled successfully. If dust was scattered all around the world by the impact, then the extraterrestrial iridium ought to show up on the other side of the globe, as far away as possible from Europe and North Africa. It duly did, in samples obtained near Canterbury, New Zealand. The best possible evidence would be the discovery of a crater of the right size (150–200 kilometres across) and the right age (sixty-five million years). There was no obvious candidate among the known craters on land, and the cosmic cannonball may have struck the sea, or a piece of land that is now submerged. For example, there is a suspicious ring-shaped feature in the seabed off the north coast of Australia. Before you rush for the map, let me mention that sixty-five million years ago the world was very different: Australia was still welded to Antarctica; the Atlantic was young and comparatively narrow; and the 'Mediterranean' was an old ocean that was being squeezed as Africa and Europe converged.

The impacting object might have been a live comet, complete with its ices and volatile material. Kenneth Hsü of Zurich argued a case for it and offered some different mechanisms for the slaughter: fierce heating of the air, poisoning of living things by cyanide brought in by the comet and spread by ocean currents, and a drastic increase in the carbon dioxide dissolved in the sea water. A seasoned investigator of cosmic impacts, Eugene Shoemaker, argued that it would be hard to account for the relative abundance of iridium, from a live comet of the right sort of size. Judged simply by the statistics of sky pollution, the odds are roughly fifty-fifty on whether so large an event would be caused by a comet or an apollo: massive apollos are much rarer than massive comets, but they blunder around for much longer. The intensive studies that the discovery is provoking in many laboratories may help to settle the issue. Some experts are not yet satisfied that the exploding-star hypothesis, which might account for both the origin of the iridium and the death of so many plants and animals, has been completely excluded. But perhaps the best argument in favour of a collision is the simplest: such an event must occur from time to time.

What would it be like, if you could witness it? To answer even the hypothetical question gives a hostage to unreason. But were it a live comet it would fill half the sky; if a dead apollo it would appear about as large as the planet Venus when half an hour off, and as wide as the Moon forty seconds before impact. With a good view of the fall you would not live to tell the tale – although you might have a moment to reflect on the sight of a blue incandescence larger than Mount Everest slanting into the atmosphere at fifty times the speed of sound, and tearing at the ground or the sea with a blinding explosion. The mushroom of debris rushing high into the air would be plainly visible as a wide, flattened cloud-top from hundreds of kilometres away. At that sort of distance the

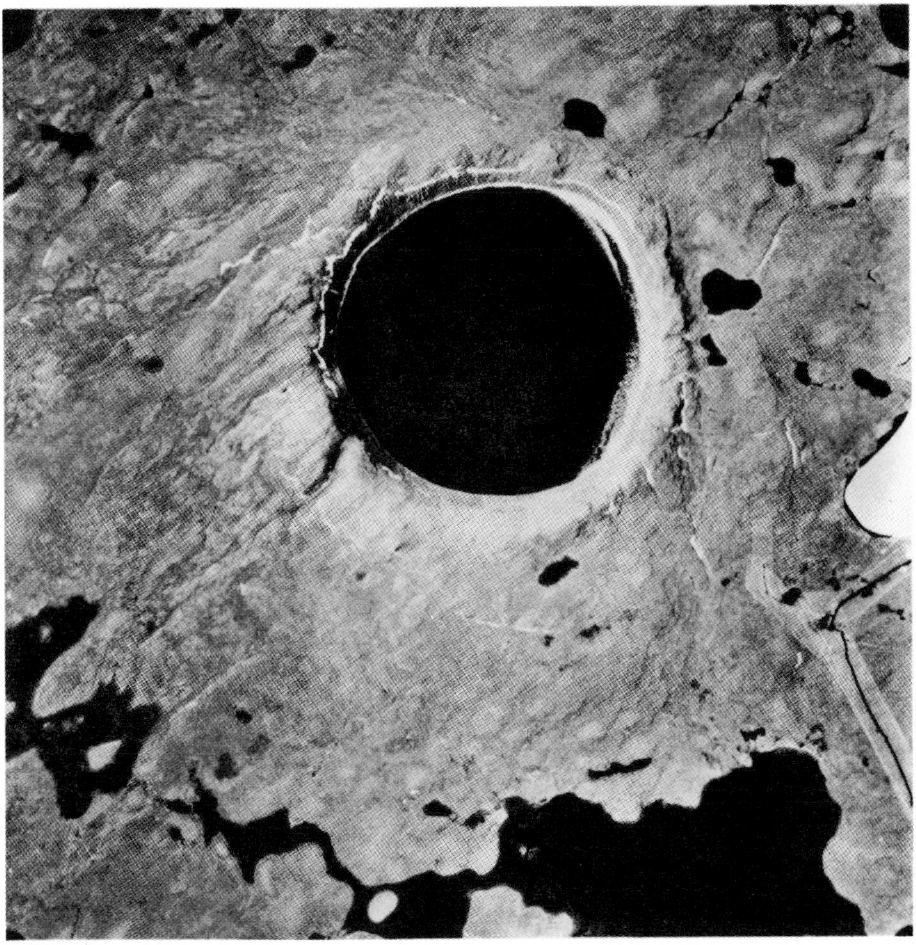

An aerial view of an impact crater, filled by a lake, New Quebec. For a much larger crater, see the colour photograph facing page 64. (R. Grieve, Canadian Department of Energy, Mines and Resources.)

blast wave would arrive in a few minutes and burst the onlooker's lungs.

The blast wave, accompanied by earthquake-like tremors, would carry the dire news all around the world and kill many animals at once. If the sea were struck, large 'tidal waves' (tsunamis) would race across the oceans and break over the coastal zones. The cloud would bring immediate darkness to the scene of impact: in remote places, the grains spreading in the stratospheric winds would at first produce ruddy sunsets and sunrises and the Moon would turn blue. One day the Sun would fail to rise.

The terminal blackout would begin within a few weeks, in a band around the world at the same latitude as the explosion. By comparison with the way radioactive dust travels in the upper atmosphere, almost a year might elapse before the whole planet was covered; as the global weather would be greatly affected this becomes speculative. But the nearest place of safety would be on the Moon, from which vantage point this planet would appear as veiled as Venus does to us.

To turn the proposition around, if the comet or apollo in question had come zipping along on its carefree way just half an hour earlier or later, it would have missed. The dinosaurs would remain in charge of the Earth and we should still be squabbling over the ants' eggs. I cannot resist teasing my friends who believe too readily in communicative aliens by citing all those millions of planets in the Milky Way now dominated by exo-dinosaurs, where the exo-shrews are waiting impatiently for the exo-comet to come, so that they can get on with their evolution, build their radio telescopes and call up the Earth.

Nevertheless, to dance a jig on the coffin of the dinosaurs would be unbecoming for any mortal who has ever felt afraid of the dark. I admit to feeling a little sorry for the dinosaurs and hoping that most of the large ones knew little about it. *Alamosaurus* of the long neck, *Triceratops* of the sportive horns and *Tyrannosaurus* of the hyperactive jaws may have been felled at one blow by the world-wide blast wave. Any that lived would face lingering starvation, deafened by the blast and with eyes unsighted in the pitch darkness.

To anyone now embracing this hypothesis, the iridium-rich clay is the most disturbing sight in the world. But there is one dust-grain of comfort. If the dinosaurs had died because of internal disorders of the planet, similar consequences might have followed, inadvertently and unpredictably, from human insults to the natural environment. The comet-or-apollo phenomenon is hard to imitate except by nuclear war, which may be considered undesirable for other reasons. Detonation of existing nuclear weapons amounting, say, to 10,000 megatons in aggregate would inject about as much persistent dust into the stratosphere as the volcano Krakatau did – considerable, chilling, but not in itself fatal. Even when blast and radiation are taken into account, it seems incapable of destroying all mammalian life. If that is the aim of the weapon-makers they will have to try harder.

The fate that overtook the dinosaurs and ammonites sixty-five million years ago was not the only event of its kind. A comparable massacre among the animals, in which nearly all perished, occur-

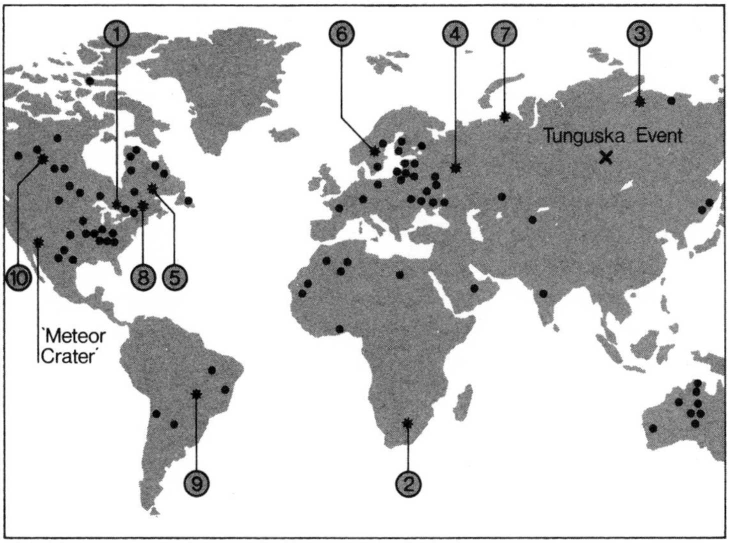

Battered Earth: the locations of craters due to known and suspected cosmic impacts, as collated by R. Grieve of Ottawa in 1979. Characteristics of the largest candidates are noted in the accompanying table. For comparison, the famous Barringer 'Meteor Crater' in Arizona (the largest from which meteoritic fragments have been recovered) is only 25,000 years old and 1.2 kilometres in diameter. Canada and the USSR have not been special targets for impacting objects; for comments on the apparently uneven distribution, see the text.

The largest suspected impact craters

No. on map	Location	Approx. age (million years)	Diameter (kilometres)
1	Sudbury, Canada	1840	140
2	Vredefort, South Africa	1970	140
3	Popigai, USSR	38	100
4	Pochezh-Katunki, USSR	183	80
5	Manicouagan, Canada	210	70
6	Siljan, Sweden	365	52
7	Karla, USSR	57	50
8	Charlevoix, Canada	360	46
9	Araguainha Dome, Brazil	less than 250	40
10	Carswell, Canada	485	37

red almost 250 million years ago, at the end of the geological period known as the Permian. The casualty lists on that occasion included the famous trilobites and sea scorpions of the sea floor, as well as many early species of reptiles. Already Smit and other geologists are investigating a tell-tale layer of clay at the top of the Permian rocks.

Lesser impacts must have been responsible for lesser disruptions of life, plausibly of the kinds associated with the transitions between geological periods in general. For example, the ammonites became conspicuous after an event, 212 million years ago, when the Triassic period gave way to the Jurassic period. A candidate crater exists at Manicougan in Canada. At Siljan in Sweden a crater fifty kilometres in diameter and dated at about 365 million years old may correlate with a minor massacre of marine invertebrates. At Popigai in Siberia there is a large crater that does not seem to coincide with any exceptional disaster among living things, if the Soviet scientists are right in saying it is 38 million years old. The evidence for extraterrestrial encounters of a deadly kind is now impressive, but the chain of cause and effect is too slack, so far, for anyone to imagine that the story is complete.

A glance at the accompanying map of suspected impact craters might suggest that Canada and the USSR attract a peculiarly large number of cosmic strikes. On the contrary, the frequency of cratering in those regions gives an impression of the spattering that must have occurred in all parts of the world. The factors that conspire to give a high 'score' in these northern countries are their very old rocks, geologically well preserved and thoroughly scoured in the ice ages, and the presence of a few Canadian and Soviet scientists who have been keenly interested in searching for craters. Most impacts must have occurred in the oceans, but the evidence is not only hard to reach, but liable to be erased every hundred million years or so by the 'recycling' of the ocean floor. While half the world is looking out for *Halley* again, the crater spotters are having a field day scanning satellite pictures of the Earth for pockmarks left by the few comets that really mattered.

8. A REMEDY FOR COMETS

Any honest investigator of sky pollution, concerned to allay comet fever rather than promote it, must say promptly and firmly that there is no risk whatever of the comet *Halley* colliding with the Earth in 1986. Its computed trajectory will keep it farther off than Venus ever comes, which is one reason why the present apparition will be unspectacular. The last time *Halley* came near was in AD 837 when it passed six million kilometres away. The closest visible comet whose distance has been measured in the past few centuries was *Lexell* of 1770, just over a million kilometres away but still well beyond the Moon. In space traffic control, a miss is as good as a megaparsec.

But astrosociology is the key to celestial interests in the mid-1980s. The first principle in this neglected branch of learning is stated most concisely by Brian Marsden: 'To the man in the street, the Solar System consists of Mars, the Rings of Saturn and Halley's Comet.' This warns us right away that comet fever in 1985–6 will be worse than at any time since 1910, when *Halley* last appeared. But the principle is also a guide to cosmic grantsmanship. When comet experts decided that they wanted to send a space probe to intercept *Halley*, its special standing as one of just three objects in the Solar System was their chief hope for unlocking the nations' coffers and accomplishing their mission. The Vikings went to Mars and the Voyagers set off for Saturn; if no one shot at *Halley* at the only opportunity for seventy-six years, the comet buffs promised that the public would demand to know the reason why.

Nevertheless, funds for space science were always tight and plans for *Halley*, drawn up with loving care by more than half a dozen working groups centred on the Jet Propulsion Laboratory in California, were delayed, abandoned and whittled down to ever slighter projects. In 1977 the intention was to launch the spacecraft in 1981–2 and have it deploy either vast solar sails, or huge solar-electric panels powering an 'ion drive' engine. Operating continuously for three to four years the propulsion system would have built up sufficient speed in the appropriate direction to enable the spacecraft to fly in company with *Halley* for several months, and to close to within a few kilometres of the nucleus, perhaps even landing on it. But first the solar sail was abandoned on the grounds that it was an unproved concept, and then the mission proposal was rejected altogether by the National Aeronautics and Space Administration, for want of time and funds.

With their deadline in the sky approaching, scientists and engineers at JPL, and their advisers from two dozen institutions, spent 1978–9 evolving a new mission to be undertaken jointly by NASA and the European Space Agency. The launch was postponed until July 1985, which made a *Halley* rendezvous in the companionable sense quite impossible. The aim was to 'fly by' in November 1985: the solar-electric ion engine would carry the main American spacecraft across *Halley*'s bows, clearing it by 130,000 kilometres to avoid physical contact. Fifteen days before that, it was to have released a small European-built probe, which would carry instruments right into the comet's head, at a relative speed of almost sixty kilometres per second. The main spacecraft would then continue on a further, more elaborate mission: to make a rendezvous with the decaying comet *Tempel 2* in 1988 and study it for a year. But that plan, too, was shot down in Washington, when the irreplaceable ion engine disappeared from the budget. Thereafter the only chance remaining for the comet scientists was to shoot a probe directly from the Earth, using conventional rocketry.

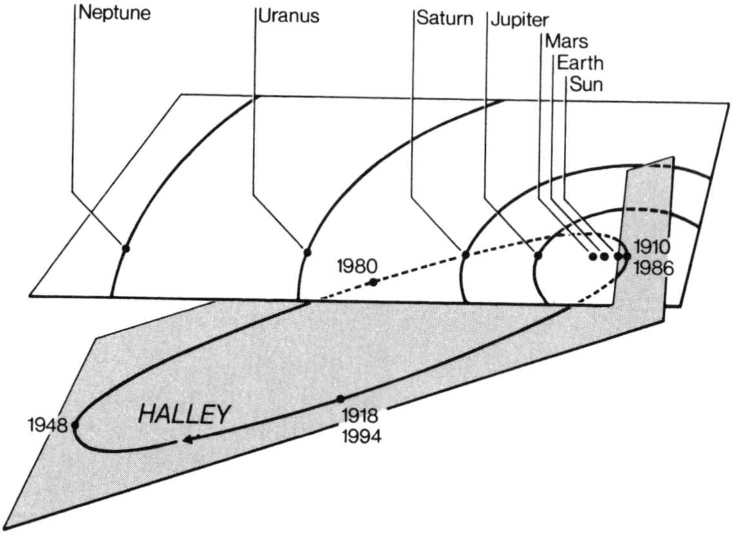

The orbit of Halley's Comet carries it far south of the plane in which the Earth and other planets revolve. The comet crosses that plane near to the Earth's orbit.

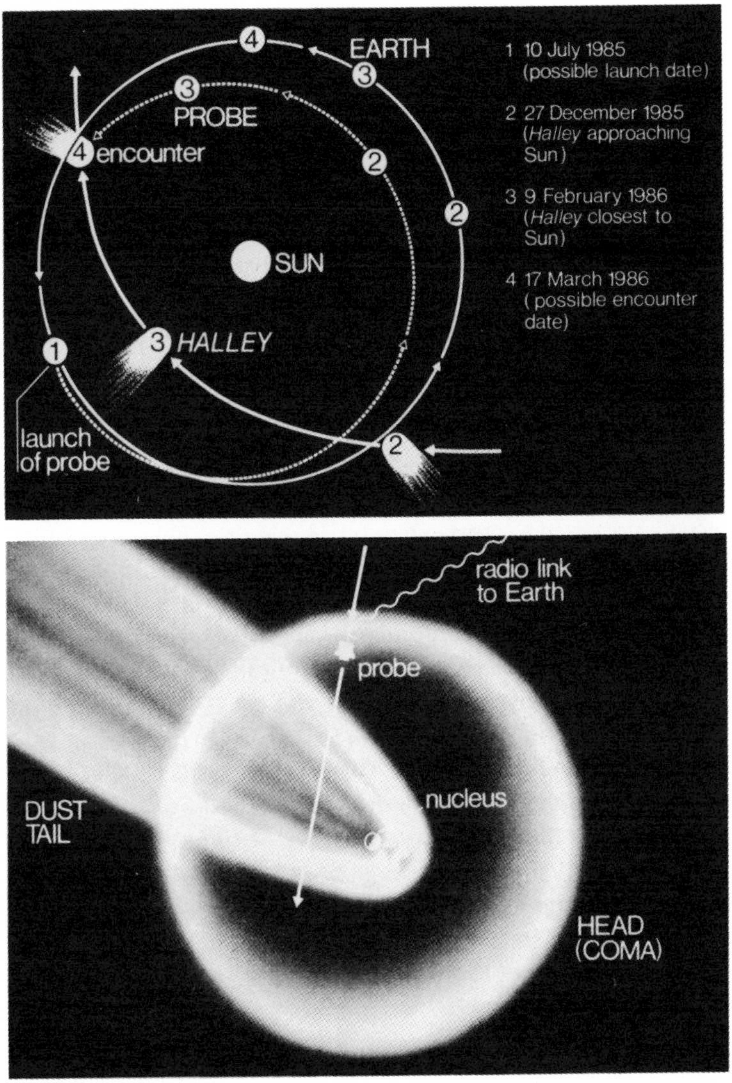

Intercepting Halley's Comet by the simplest method means (1) launching a probe long before the intended encounter and (2) letting the probe plunge into the comet at a very high relative speed. These diagrams illustrate the European Space Agency's Giotto mission authorised in 1980.

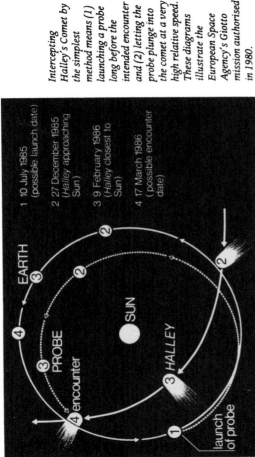

The easiest trajectory for a spacecraft to *Halley* is one that aims to reach the comet when it is just crossing the plane of the Earth's orbit. Even so, an encounter of that kind with *Halley* in the spring of 1986, when the comet is heading outwards after visiting the Sun, still requires a launch no later than the summer of 1985. A Soviet project for sending spacecraft into orbit around Venus (and then to release French-made balloons into the planet's atmosphere) is being modified to send one of the spacecraft onwards, to the comet, aimed to pass less than 50,000 kilometres from the nucleus. The Japanese hope to observe *Halley*, though more remotely, with a space-borne ultraviolet telescope. The Americans turned their attention to a probe to be launched from the Space Shuttle, to intercept the comet. But the first shot at *Halley* officially confirmed (in July 1980) was the European Space Agency's mission. It will close to within 1000 kilometres of the comet.

Named Giotto after the comet's portraitist mentioned in Chapter 1, this probe will be despatched from the Earth by an Ariane launcher, on behalf of the countries of Western Europe, most of which participate in ESA. The spacecraft will weigh about 750 kilograms (three-quarters of a ton) at launch and during its eight-month journey it will observe the comet's ultraviolet emissions. If all goes well, the adventure will culminate in a frenzy of activity as Giotto plunges into the dust of *Halley* at 68 kilometres per second, protected to some extent by double-skin 'bumpers', while a spinning camera transmits pictures of the inner regions of the comet's head, for observers to peer at, looking for the nucleus.

A simple *Halley* probe, flying through the comet's head at a very high relative speed, has been aptly called a kamikaze mission. Shotblasted by comet dust, the spacecraft will be lucky to survive for the hour or so needed to travel close to the nucleus and obtain pictures of the snowball. Other experiments on board will study the dust itself and the ionised gas, while the scientists back on Earth keep their fingers crossed that their instruments will escape destruction for long enough to gather useful results. But it has not been judged necessary to sterilise the probe, to protect the alleged influenza factories of *Halley* from earthly contamination.

In the International Halley Watch, the 'deep space' missions are co-ordinated with observations from spacecraft in Earth orbit and from ground-based observatories all around the world. Since *Halley*'s last visit in 1910, stargazing has changed almost beyond recognition. At that time astronomers prided themselves that they could not only watch the comet through their telescopes but also take photographs of it and analyse its light with spectroscopes. Then, the most powerful telescope was 1½ metres (60 inches) wide, at Mount Wilson. To say that in 1985–6 there will be sixteen optical telescopes between 2½ and six metres wide, scattered around the globe, and one of 2.4 metres going into space, does not convey the enormous enhancement of their power, in an era when individual particles of light can be recorded by electronic detectors. Besides tracking the comet coming and going at far greater distances than ever before, the telescopes may, at closer range, pick out details not previously recorded. A large number of lesser but very sophisticated instruments, including wide-angle Schmidt telescopes, will take a major share of the ground-based studies of *Halley*, using visible light.

But that is only the beginning. Since 1945 a succession of 'new astronomies' have introduced radio telescopes and infrared telescopes on the ground, and also instruments in orbit for detecting rays from the universe that do not penetrate the Earth's atmosphere; these range from long radio waves to energetic X-rays and gamma rays. Many of the new ground-based and space-based techniques were tried out on *Kohoutek* in the early 1970s, with results that gratified the experts. They thus have every opportunity, with infrared and ultraviolet observations in particular, for revealing features and compositions not to be seen with visible light.

In their efforts to unmask *Halley* the professional astronomers and space scientists will lose a lot of sleep, but probably less than the amateur astronomers. For many of these, with no comets to their names, the apparition will be the event of a lifetime and they will take upon themselves the solemn duty of watching over *Halley* night after night, whenever it is visible. Only friendly clouds and the Earth's horizon, blotting out the view, will stand between the amateurs and the divorce courts. While they cannot begin to compete with the professionals, one gambler's chance will comfort them better than thermos coffee: they might see the comet carrying out the threat it allegedly made in 1910, and breaking up into two or more pieces.

If *Halley* clones, heaven help us: who will check the excitement then? And if the splitting should occur after a probe has gone through the comet no one will listen to careful explanations about how it passed too far from the nucleus to do any damage. The space scientists will be arraigned for cosmic vandalism. But the odds are against a break-up and in any case, when *Halley* rounds the Sun on 9 February 1986, it will be on almost exactly the opposite side from the Earth. Astronomers and members of the public in the northern hemisphere will simply find the comet hard to see.

Donald Yeomans of JPL, who has estimated the comet's brightness from stage to stage as well as its changing positions in the sky, suggests that when *Halley* passes the Earth on the way in, during

Donald Yeomans with his computer print-out giving the track (ephemeris) of Halley's Comet across the sky at the forthcoming apparition. (JPL.)

November 1985, it will be invisible to the naked eye. At the end of December a good pair of binoculars may serve to pick it out, after sunset. It may become visible to the unaided eye in the third week of January 1986, low on the western horizon after sunset, but no brighter than a relatively faint star. The comet will then pass beyond the Sun, a little to the north of it, and it will be lost in the bright sunlight.

Once on the other side of the Sun, *Halley* will be in view from the Earth before dawn. At the end of February it may be accessible to naked eyes an hour before sunrise. After that the comet will rise earlier each day, while sidling away towards the south. On 7 March it will be almost in line with the Moon. By mid-March the tail, if you can make it out at all, may appear to stretch across one-sixth of the sky, or thereabouts.

On 11 April, when *Halley* is at its closest to the Earth, it will be high overhead in Australia, New Zealand and Argentina, in the constellation of Lupus, and relatively bright; but at that time it will be very low on the southern horizon from the USA and the Mediterranean and invisible from northern Europe. (No doubt chartered jets and cruise liners will take enthusiasts south of the equator in April, to observe the comet at its meagre best.)

Then the comet will creep north again, as an evening object, but growing fainter all the time. The best view from the northern hemisphere may offer itself during the last few days of April. In early May there will be a shower of meteors (shooting stars) as the Earth comes close to *Halley*'s orbit, seven weeks after the comet itself has passed the spot. But to watch the comet itself steaming away in May will require powerful binoculars or a small telescope.

All of these forecasts of visibility assume clear and dark skies. Electric lights ensure that, over a typical modern city, the sky is as bright as it is in rural areas at the time of Full Moon. Few city-dwellers are likely to see *Halley* with their own eyes, this time around. But there will be a stream of pictures originating from the professional observatories and spacecraft. At this apparition *Halley* makes its debut on television.

Even if it is outshone by a more brilliant, unexpected comet, as happened in 1910, many people will fail to distinguish between them. *Halley* is *Halley*, item number three in the Solar System. Sir Bernard Lovell and other astronomers have already proclaimed this unsatisfactory fireworks show as the cosmic event of the decade. The fever is upon us.

The newspaper headlines are as predictable as the comet's path: 'Halley pleases astronomers, not public' and 'Do path doubts mean comet may strike?' (Answer, 'No', in much smaller print.) And, of course, to caption a photograph: 'Halleylujah!' There will be Halley silver medals, Halley posters, comet cocktails and buttons saying 'Halley's Cometh'. Despite modern drugs legislation, the grandsons of the quacks who proffered comet pills in 1910 will take several leaves out of Hoyle and Wickramasinghe's *Diseases from Space* and find ways of marketing protections against everything from the common cold to yellow fever. (They will ignore the inconvenient dissociation, even in that hypothesis, between the time of apparition and the arrival of diseases.) The enterprising folk who sold helmets in 1979, when the space station Skylab was making its long-heralded descent to the Earth, will dust off their old stocks and sell some more as comet-proof headwear.

Halley will be less than a great comet this time, but the astrologers will be unabashed in interpreting it as a sign of all kinds of disasters. They will not lightly give up the prerogatives of thousands of years, when there is a well-publicised comet around; the usual pathetic crop of suicides will no doubt ensue, among impressionable people, and who knows what superstitious military governments may be tempted to imitate Nero, and bump off the opposition? Those citizens who believe that Uri Geller bends spoons using occult powers are likely also to suppose that the comet portends dreadful events. At the other end of the scale, and probably completely immune to the fever, is a man who thinks that a comet is an obsolescent jet aircraft, and who works underground in an Arctic mine on the night shift.

The prediction that *Halley* will miss us by sixty million kilometres on 11 April 1986 will scarcely interfere with the inevitable prophecies of a collision. I eavesdropped once on two small boys who had cycled out to the local airport to watch the big jets crashing. As each aircraft took off or touched down they reported to each other, with just a hint of disappointment: 'That one didn't crash.' I imagine them both in 1986, older but no wiser, saying to their wives as they watch Halley's Comet fading into the outer darkness: 'That one didn't crash.'

The final and specific remedy for sky pollution and comet fever will come with practical measures of comet control. From all the centuries of nonsense, the only substantial reason why anyone should give more than a glance to passing comets is the fear of collision. Although the risks of a really big impact by a comet or apollo during, say, the next million years are slight, the environmental consequences would be regrettable. For this reason nature lovers (and even some people lovers) begin to speak in terms of planetary conservation.

In Project Icarus at the Massachusetts Institute of Technology in the late 1960s, students devised a scheme for warding off a putative strike by an apollo one kilometre in diameter – the sort of event expected once in 250,000 years. They envisage using Saturn V moon-rockets, launched in the last thirteen days before impact, to deliver six 100-megaton H-bombs that would fragment or deflect the apollo. That was an improvised scheme, and in future an international sky-cleaning service will no doubt maintain a nuclear umbrella around the planet, against wayward comets and apollos. It will have telescopes and radars, based in space as well as on the ground, devoted full time to tracking the pollution and using whatever powerful weapons may be necessary to intercept a menacing object and render it harmless. While they wait for a substantial target – perhaps for a thousand careers – they can practise on smaller meteorites and inconsequential comets, although I understand that proposals for bombing *Halley* at the present apparition have not found favour.

Here at last is a practical use for all that military and space hardware that has accumulated in recent decades: to let young lads from Boise to Boston, and old ladies from North Cape to South Island, sleep soundly in their beds, untroubled by thoughts of comets landing in the potato patch. They will know that the brave Space Scouts are out there, ready to blast any sneaky intruders right back to the Öoo Cloud where they belong. Amateur astronomers will find a new purpose in life, supporting the space patrols from the rooftops.

Not that we need sit passively and wait for the comets and apollos to threaten us. Freeman Dyson suggests that *Mayflowers* of the space age will take people to colonise the nearer microplanets. And some American astronomers want to send out space tugs that would bring back selected apollos and put them into orbit around the Earth. Then they would be quarried for useful minerals and metals, and the products would either go into building cities in space or be splashed in wing-shaped parcels into the oceans, for recovery and use by earthlings. But I can't help picturing the tug skipper arriving home with his apollo and making a slight error of navigation, in an operation that is the space equivalent of reversing a car into a garage. Oops! a new geological period begins.

George Wetherill may be right in wanting to declare the apollos a wilderness area. Comets too have their supporters anxious to fence them off from human interference, despite (or even because of) their ways of smearing the sky. But if, of its own accord, one of these national parks in solar orbit is reckoned to be heading our way, self-preservation presumably comes first. All life on Earth is a commonwealth unconsciously maintaining the conditions necessary for survival – or so James Lovelock has suggested in his hypothesis of Gaia – and our function as the allegedly intelligent species becomes apparent. Earth in her maternal wisdom has spawned us to use our wits to protect all living things from the Curse of the Dinosaurs.

In this otherwise inspiring project, I note some drawbacks. It will make evolution more easy-going, not to say reactionary. If the army ants, for instance, are expecting the next big bang to help them wrest from us the domination of the planet, they will gnash their mandibles in vain. Among humans, on the other hand, the Space Scoutmaster, orbiting high above us and armed to the teeth, will be well placed to claim mastery of the world. Perhaps the weightiest objection of all is this: if Lovelock's fancy is valid then the only technology that matters would seem to be the making of rocket-powered H-bombs, while the comet hunter stands on a plinth, at the pinnacle of all evolution.

Fortunately we need be in no great hurry to set up the anti-comet batteries. In a world where the nuclear weapons are ready on the instant to defend us against one another by blowing us all up, to fret about cosmic impacts is like worrying about being struck by lightning during the Battle of the Somme. But human beings have only a vague sense of probabilities, and to fear the fall of a comet is no longer entirely irrational, if you say it may be a dead one and point to apollos in the offing like Betulia, six kilometres, and 1978 SB, eight kilometres in diameter. By the time *Halley* comes round again in 2061, precautions against comets will seem as natural as keeping down mice.

Scientists and others who wish to rally support for this large-scale redeployment of resources should remember Chicken Licken. When the acorn fell on his head, in the old story, he realised at once what was happening, and set off with all the local poultry to report to the king that the sky was falling down. A family of foxes aborted their mission by eating them. Quite the wrong moral is often drawn from this little tragedy: it is not that you should curb your hypothesis-making, but to be wary of foxes who only pretend to believe you.

Albrecht Dürer used a comet as a symbol
of melancholia in his famous engraving.

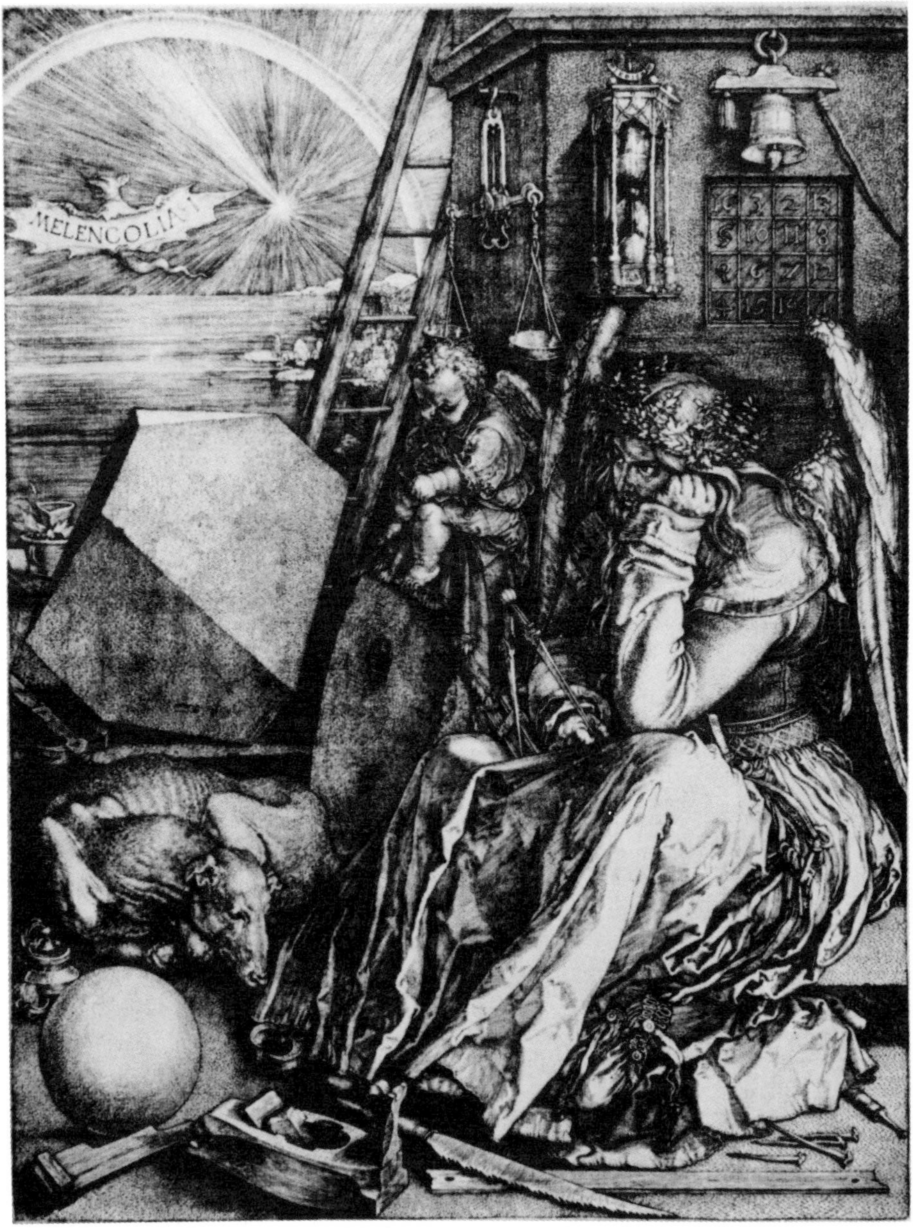

BIBLIOGRAPHY

Further reading

This short list is confined to a selection of works accessible (in both senses of the word) to the general reader. More specific textual and technical references follow.

For recent histories of cometary superstitions and science:

Peter Lancaster Brown *Comets, Meteorites and Men* Robert Hale (1973).
Patrick Moore *Guide to Comets* Lutterworth (1977).
Philippe Véron and Jean-Claude Ribes *Les Comètes* Hachette (1979) *in French*.

For the life of Edmond Halley:

Angus Armitage *Edmond Halley* Nelson (1966).
Colin A. Ronan *Edmond Halley: Genius in Eclipse* Doubleday (1969).

For fairly simple accounts of modern comet science:

Fred Whipple 'The Spin of Comets' in *Scientific American* March 1980.
(Ed.) Marcia Neugebauer and others *Space Missions to Comets* NASA CP-2089 (1979). This is simpler than, for example, (ed.) B. Donn and others *The Study of Comets* (two parts) NASA SP-343 (1976).

For the hypothesis of comet-borne diseases:

Fred Hoyle & N. C. Wickramasinghe *Diseases from Space* Dent (1979).

For cosmic impacts:

George Wetherill 'Apollo Objects' in *Scientific American* March 1979.
Luis Alvarez and others: 'Extraterrestrial Cause for the Cretaceous-Tertiary Extinction' in *Science* Vol. 208 p. 1095 (6 June 1980). (This paper is technical, but not quite impossible for the layman. The papers by Jan Smit & Jan Hertogen and by Kenneth Hsü, given in the technical references, are more compressed and correspondingly harder for non-scientists to follow.)

Some textual and technical references

p. 10 Swindle: 'I think that comets are the greatest little deceivers in the Solar System.' (F. Whipple 1977.)
p. 11 M. Dubin: foreword to (ed.) G. A. Gary *Comet Kohoutek* NASA SP-355 (1975).
p. 12 W. Shakespeare *Henry VI Part 1*, I,i.
p. 12 Tacitus *Annals* (trs. M. Grant) Penguin (1956).
p. 13 Suetonius *The Twelve Caesars* (trs. R. Graves) Penguin (1957).
p. 13 Sacred Followers: cited in P. Véron and J.-C. Ribes *Les Comètes* Hachette (1979).
p. 13 Charlemagne's non-comet: A. Pingré *Cometographie* Vol. 1 (1783). In the Library of the Royal Astronomical Society, London.
p. 14 J. Needham *Science and Civilisation in China* Vol. 3 Cambridge U.P. (1959).
p. 16 'Guest star' 1408: see *Sky and Telescope* Vol. 58 p. 323 (1979).

p. 16 Clockwork: see, for example, J. H. Plumb *In the Light of History* Allen Lane (1972).
p. 17 'Black holes have no hair': see, for example, C. Misner, K. Thorne & J. A. Wheeler *Gravitation* Freeman (1973).
p. 17 King Wu: quoted by Y. C. Chang in *Acta Astronomica Sinica* Vol. 19 p. 109 (1978) trs. into English in *Chinese Astronomy* Vol. 3 p. 120 (1979).
p. 17 J. Needham as p. 14.
p. 18 Chang as p. 17.
p. 18 D. Yeomans, personal communication. See also T. Kiang in *Memoirs of the Royal Astronomical Society* Vol. 76 p. 728 (1971).
p. 20 Giotto's painting: R. Olson in *Scientific American* May 1979.
p. 21 D. Hughes *The Star of Bethlehem Mystery* Dent (1979).
p. 21 P. Toscanelli: in G. Uzielli & G. Celoria *Paolo dal Pozzo Toscanelli* (1894).
p. 22 T. More *Utopia* (trs. P. Turner) Penguin (1965).
p. 22 W. Shakespeare *King Lear* I,ii.
p. 23 *Cometomantia* (1684). In the Library of the Royal Astronomical Society, London.
p. 24 Vagrancy Act 1824.
p. 25 M. David *Forty Days! and Ninevah Shall Be Destroyed!* Children of God Trust (1973).
p. 25 'French writer': Camille Flammarion (1909), cited in P. Véron & J.-C. Ribes *Les Comètes*, Hachette (1979).
p. 26 J. Thurber *My World and Welcome To It* Hamish Hamilton (1942).
p. 27 E. Halley biographies: see 'Further reading'; also *Dictionary of National Biography*.
p. 28 Aristotle *Meteorology* Book 1.
p. 30 Tycho Brahe incident in, for example, J. Dreyer *Tycho Brahe* (1890).
p. 32 Tycho Brahe *De Nova Stella* (1573) trs., for example, in *A Treasury of World Science* Philosophical Library (1962).
p. 32 M. Schmidt in *The Violent Universe* television production (Philip Daly & Nigel Calder) BBC-TV 1969.
p. 33 Tycho Brahe as p. 32.
p. 33 Tycho and the comet of 1577: see, for example, Owen Gingerich *Sky and Telescope* Vol. 54 p. 452 (1977).
p. 34 W. Shakespeare *Troilus and Cressida* I,iii.
p. 35 Pope John Paul II: address trs. in *Science* Vol. 207 p. 1165 (1980) also *International Herald Tribune* 11 Nov 1979.
p. 35 'A banterer of religion': see, for example, W. Whiston *Memoirs* (1749), cited in S. Schaffer *Notes and Records of the Royal Society* Vol. 32 p. 17 (1977).
p. 35 Kepler's planetary tunes: for example, hear J. Rodgers & W. Ruff *The Harmony of the World* (gramophone record) Sky Publishing (1979).
p. 36 J. Kepler *De Cometis* (1619) and G. Galileo *Il Saggiatore* (1625) cited in Ruffner (next entry).
p. 37 J. Ruffner in *Journal for the History of Astronomy* Vol. 2 p. 178 (1974).
p. 38 J. Hevelius *Cometographia* (1668). In the Library of Royal Astronomical Society, London.
p. 39 E. Halley 'Ode to Newton' trs. L. Richardson in Newton's *Principia* U. California Press (1934).
p. 40 I. Newton *Principia* (1687 and 1713) trs. A. Motte (1729) rev. F. Cajori (1934) U. California Press (1934).
p. 41 Newton's breakdown: see, for example, 'Science and the Citizen' in *Scientific American* December 1979.
p. 41 Halley's brandy: J. Flamsteed's letter of 18 December 1703: 'Dr Wallis is dead: Mr Halley expects his place, who now talks, swears, and drinks

brandy like a sea-captain: so that I fear his own ill behaviour will deprive him of the advantage of this vacancy.'

p. 43 E. Halley *Philosophical Transactions of the Royal Society* Vol. 24 p. 1882 (1705).

p. 43 E. Halley's 'late note': in *Astronomical Tables* (1752).

p. 44 C. Ronan *Edmond Halley* Doubleday (1969).

p. 44 Bellman in Lewis Carroll *The Hunting of the Snark* (1876).

p. 44 The planet Vulcan: see, for example, R. Gould *Oddities* Philip Allan (1928).

p. 45 Newtonian 'horrible sight': see, for example, N. Calder *Einstein's Universe* BBC and Viking (1979).

p. 47 Comet names in B. Marsden *Catalogue of Cometary Orbits* 3rd ed. (1979).

p. 50 P. Lancaster Brown *Comets, Meteorites and Men* Robert Hale (1973).

p. 52 B. Marsden *The Central Bureau for Astronomical Telegrams* Leaflet No. 493 of the Astronomical Society of the Pacific (1970).

p. 54 A. Delsemme, personal communication. See also (ed.) A. Delsemme *Comets, Asteroids, Meteorites* U. Toledo (1977).

p. 57 S. Vsekhsvyatsky *Physical Characteristics of Comets* Moscow (1958) *in Russian* trs. NASA (1964).

p. 57 T. Van Flandern, personal communication.

p. 57 W. Napier & V. Clube in *Nature* Vol. 282 p. 455 (1979).

p. 59 E. Öpik in *Proceedings of the American Academy of Arts and Sciences* Vol. 67 p. 169 (1932).

p. 59 J. Oort in *Bulletin of the Astronomical Institutes of the Netherlands* Vol. 11 p. 91 (1950).

p. 60 E. Halley *Philosophical Transactions of the Royal Society* Vol. 30 p. 736 (1718).

p. 63 Bombardment: see, for example, G. Wetherill in *Proceedings of the Sixth Lunar Science Conference* p. 1539 (1975).

p. 65 J. Bruner, personal communication. Also *Inside Out, Outside In* in press.

p. 65 T. Kuhn *The Structure of Scientific Revolutions* 2nd Edition U. Chicago Press (1970).

p. 66 A. Paré quoted, for example, in P. Véron & J.-C. Ribes *Les Comètes* Hachette (1979).

p. 66 *Cometomantia* as p. 24.

p. 66 Seneca *Questiones Naturales.*

p. 66 Challis: see, for example, R. Lyttleton *The Comets and their Origin* Cambridge U.P. (1953).

p. 67 Z. Sekanina in *Icarus* Vol. 38 p. 300 (1979).

p. 68 B. Marsden, personal communication.

p. 68 Martian canals: see, for example, C. Sagan *The Cosmic Connection* Doubleday (1973).

p. 69 Whirly Hell and eighteenth-century attitudes: S. Schaffer, personal communication.

p. 69 I. Newton *Principia* as p. 39.

p. 70 W. Herschel: see, for example, A. Armitage *William Herschel* Nelson (1962) and S. Schaffer in *Journal for the History of Astronomy* in press (1980).

p. 71 J. Michell in *Philosophical Transactions of the Royal Society* Vol. 74 p. 35 (1784) cited in S. Schaffer in *Journal for the History of Astronomy* Vol. 10 p. 42 (1979).

p. 72 I. Newton *Principia* as p. 39.

p. 74 J. Kepler *De Cometis* (1619).

p. 78 L. Biermann in *Zeitschrift für Astrophysik* Vol. 29 p. 274 (1951).

p. 80 P. Lancaster Brown as p. 50.
p. 81 Soviet comet simulation: I. Podgorny and others in *Astrophysics and Space Science* Vol. 61 p. 369 (1979).
p. 85 D. Yeomans' ephemeris of Halley's Comet in Appendix B of *A First Comet Mission* NASA TM-78420 (1977). See also D. Yeomans in *Astronomical Journal* Vol. 82 p. 435 (1977).
p. 86 H. Russell quoted in R. Lyttleton (next entry).
p. 87 R. Lyttleton *The Comets and their Origin* Cambridge U.P. (1953).
p. 87 Voltaire *Micromégas* (1752).
p. 88 F. Whipple in *Astrophysical Journal* Vol. 111 p. 375 (1950) and Vol. 113 p. 464 (1951).
p. 90 Encke snowball: see, for example, B. Marsden & Z. Sekanina in *Astronomical Journal* Vol. 79 p. 413 (1974) and F. Whipple 'The Spin of Comets' in *Scientific American* March 1980.
p. 91 R. Newburn and D. Yeomans in Appendix A of *A First Comet Mission* NASA TM-78420 (1977). See also R. Newburn in *The Comet Halley Micrometeoroid Hazard* European Space Agency SP-153 (1979).
p. 92 *Kohoutek*: (ed.) G. Gary *Comet Kohoutek* NASA SP-355 (1975).
p. 95 D. Brownlee and others in (ed.) A. Delsemme *Comets, Asteroids, Meteorites* U. Toledo (1977).
p. 96 Meteorites from comets: G. Wetherill in *Geochimica and Cosmochimica Acta* Vol. 40 p. 1297 (1976). For arguments against the comet-apollo link see D. Hughes, *Nature* Vol. 286 p. 10 (1980).
p. 99 'Cornell astronomers': J. Gradie and J. Veverka in *Nature* Vol. 283 p. 840 (1980).
p. 99 Meteorite inclusions: see, for example, R. Clayton and others *Science* Vol. 182 p. 485 (1973) and R. Hutchinson *Nature* Vol. 283 p. 813 (1980).
p. 100 R. A. Lyttleton in *Quarterly Journal of the Royal Astronomical Society* Vol. 18 p. 213 (1977).
p. 103 Meteorite hoax: E. Anders and others in *Science* Vol. 146 p. 115 (1964).
p. 104 F. Dyson: 'The World, the Flesh and the Devil' *Bernal Lecture* (1972).
p. 105 F. Hoyle & N. C. Wickramasinghe *Diseases from Space* Dent (1979). See also (same authors) *Lifecloud* Dent (1978).
p. 105 F. Hoyle *Astronomy and Cosmology* W. H. Freeman (1975).
p. 105 Radioactivity in comets: W. Irvine and others in *Nature* Vol. 283 p. 748 (1980).
p. 106 C. Darwin in F. Darwin *Life and Letters of Charles Darwin* Vol. 3 (1887).
p. 106 D. Defoe *Journal of the Plague Year* (1722) Penguin (1966).
p. 113 'Prove Fred wrong': see, for example, N. Calder *Violent Universe* BBC and Viking (1969).
p. 116 P. Medawar *Advice to a Young Scientist* Harper & Row (1979).
p. 116 'LGM': A. Hewish, personal communication (1968).
p. 117 Manuscripts experiment: D. Peters & S. Ceci, reported in *New Scientist* Vol. 85 p. 950 (1980).
p. 118 A. Tennyson 'On a Spiteful Letter' (1868) in *Poems of Tennyson* Oxford U.P. (1913).
p. 119 E. Halley at the Royal Society 1694: published in *Philosophical Transactions* Vol. 33 p. 118 (1724).
p. 121 W. Whiston *A New Theory of the Earth* (1696). In the Library of the Royal Astronomical Society, London.
p. 122 Voltaire *Letters on England* (trs. L. Tancock) Penguin (1980).
p. 123 I. Velikovsky *Worlds in Collision* Doubleday (1950).
p. 123 J. Bedford *Will the Great Comet now Rapidly Approaching Strike the Earth?* (1857).

p. 123 L. Niven & J. Pournelle *Lucifer's Hammer* Futura (1978).

p. 123 *Meteor* a Sandy Howard/Gabriel Katzka/Sir Run Run Shaw motion picture (1979).

p. 123 G. Griffith *Olga Romanoff* (1894).

p. 123 H. G. Wells *In the Days of the Comet* (1906).

p. 125 F. J. W. Whipple in *Quarterly Journal of the Royal Meteorological Society* Vol. 16 p. 287 (1930).

p. 125 A. Ben-Manahem in *Physics of the Earth and Planetary Interiors* Vol. 11 p. 1 (1975).

p. 125 D. Hughes in *Nature* Vol. 259 p. 626 (1976) and J. Brown & D. Hughes in *Nature* Vol. 26 p. 512 (1977).

p. 126 L. Kresak: see *Sky and Telescope* Vol. 57 p. 497 (1978).

p. 126 Streamers on the Moon: P. Schultz & L. Srnka in *Nature* Vol. 284 p. 22 (1980).

p. 127 E. Öpik in *Proceedings of the Royal Irish Academy* Section A, Vol. 54 p. 165 (1951).

p. 127 G. Wetherill in *Icarus* Vol. 37 p. 96 (1979).

p. 128 E. Halley on plants: *Biographia Britannica* (1757) cited in A. Armitage *Edmond Halley* Nelson (1966).

p. 128 J. Swift *Gulliver's Travels* 'A Voyage to Laputa' (1726).

p. 133 J. Smit in *Proceedings of the Cretaceous – Tertiary Boundary Events Symposium* Vol. 2. p. 156 (1979).

p. 133 L. Alvarez and others in *Science* Vol. 208 p. 1095 (1980).

p. 133 J. Smit & J. Hertogen in *Nature* Vol. 285 p. 198 (1980).

p. 136 K. Hsü in *Nature* Vol. 285 p. 201 (1980); see also Napier and Clube as p. 57.

p. 139 Cratering record: R. Grieve & P. Robertson in *Icarus* Vol. 38 p. 212 (1979).

p. 142 Defunct mission: *Halley Flyby/Tempel 2 Rendezvous: Mission Summary* JPL 79-4 (1979).

p. 142 Probes to Halley's Comet: see, for example, *Giotto: Comet Halley Flyby* European Space Agency SCI (80) 4 (1980).

p. 144 B. Marsden in (ed.) M. Neugebauer and others *Space Missions to Comets* NASA CP 2089 (1979).

p. 145 D. Yeomans as p. 85.

p. 148 MIT Student Project in Systems Engineering *Project Icarus* MIT Press (1968).

p. 149 F. Dyson *Disturbing the Universe* Harper & Row (1979).

p. 149 Asteroid harvesting: see, for example, N. Calder *Spaceships of the Mind* BBC and Viking (1978).

p. 149 G. Wetherill 'Apollo Objects' in *Scientific American* March 1979.

p. 149 Lovelock *Gaia: A New Look at Life on Earth* Oxford U.P. (1979).

PICTURE CREDITS

Picture research: Naomi Narod

Page 8 Ann Ronan Picture Library; 14 Purple Mountain Observatory; 15 Professor Il-Seong Nha, Yonsei University, Seoul; facing 16 top Michael Holford, bottom Photographie Giraudon; facing 17 photo Derrick Witty; 18 Michael Holford; 19 Science Photo Library, London; 20 Biblioteca Nazionale Centrale, Florence/photo Pineider; 22 photo Derrick Witty; 23 Photographie Giraudon; 25 Children of God Trust 1973; 29 National Portrait Gallery, London; 30 RAS/photo Derrick Witty; 31 photo Derrick Witty; facing 32 Jet Propulsion Laboratory; facing 33, 34 Ann Ronan Picture Library; 36, 37, 40 RAS/photo Derrick Witty; 41, 46 Ann Ronan Picture Library; 48 Lick Observatory; facing 48 Space Frontiers Limited; facing 49 C. Niccolier; 51 Alan McLure, Frazier Mountain; 52 Philip Daly; 53 Australian Information Service; 55 Science Photo Library, London; 56 Yerkes Observatory; facing 64 top Earth Physics Branch, Department of Energy, Mines and Resources, Ottawa, Canada, bottom J. Smit; facing 65 Space Frontiers Limited; 67 Ann Ronan Picture Library; 73 top New Mexico State University Observatory, bottom Z. Sekanina and J. A. Farrell; 75 photo Derrick Witty; 77 top left Armagh Observatory, top right R. Fogelquist, bottom Lick Observatory; 79 Yerkes Observatory; 82 Royal Observatory, Edinburgh; 84 Helwan Observatory; 88 Smithsonian Astrophysical Observatory; 89 Yerkes Observatory; 92 D. Brownlee, University of Washington; 94 Courtesy of American Meteorite Laboratory, Denver, Colorado; 97 Space Frontiers Limited; 98 Smithsonian Institution; 107 Ann Ronan Picture Library; 108 Geoff Howard; 110 Science Photo Library, London; 118 BBC Hulton Picture Library; 120 American Museum of Natural History; 121 Ronald Grant Archive; 124 Novosti Press Agency; 129 NASA; 130 Palomar Observatory; 132 W. Alvarez, University of California, Berkeley; 137 R. Grieves, Department of Energy, Mines and Resources, Ottawa, Canada; 146 JPL; 151 British Museum, London.

Diagrams by Diagram.

INDEX

Page numbers in italics refer to captions for illustrations; f.p. = facing page

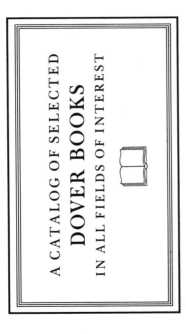

A CATALOG OF SELECTED

DOVER BOOKS

IN ALL FIELDS OF INTEREST

A CATALOG OF SELECTED

DOVER BOOKS

IN ALL FIELDS OF INTEREST

SUNDIALS, Albert Waugh. Far and away the best, most thorough coverage of ideas, mathematics concerned, types, construction, adjusting anywhere. Over 100 illustrations. 230pp. 5⅜ × 8½. 22947-5 Pa. $5.95

PICTURE HISTORY OF THE NORMANDIE: With 190 Illustrations, Frank O. Braynard. Full story of legendary French ocean liner: Art Deco interiors, design innovations, furnishings, celebrities, maiden voyage, tragic fire, much more. Extensive text. 144pp. 8⅜ × 11¾. 25257-4 Pa. $11.95

THE FIRST AMERICAN COOKBOOK: A Facsimile of "American Cookery," 1796, Amelia Simmons. Facsimile of the first American-written cookbook published in the United States contains authentic recipes for colonial favorites—pumpkin pudding, winter squash pudding, spruce beer, Indian slapjacks, and more. Introductory Essay and Glossary of colonial cooking terms. 80pp. 5⅜ × 8½. 24710-4 Pa. $3.50

101 PUZZLES IN THOUGHT AND LOGIC, C. R. Wylie, Jr. Solve murders and robberies, find out which fishermen are liars, how a blind man could possibly identify a color—purely by your own reasoning! 107pp. 5⅜ × 8½. 20367-0 Pa. $2.95

ANCIENT EGYPTIAN MYTHS AND LEGENDS, Lewis Spence. Examines animism, totemism, fetishism, creation myths, deities, alchemy, art and magic, other topics. Over 50 illustrations. 432pp. 5⅜ × 8½. 26525-0 Pa. $8.95

ANTHROPOLOGY AND MODERN LIFE, Franz Boas. Great anthropologist's classic treatise on race and culture. Introduction by Ruth Bunzel. Only inexpensive paperback edition. 255pp. 5⅜ × 8½. 25245-0 Pa. $7.95

THE TALE OF PETER RABBIT, Beatrix Potter. The inimitable Peter's terrifying adventure in Mr. McGregor's garden, with all 27 wonderful, full-color Potter illustrations. 55pp. 4¼ × 5½. 22827-4 Pa. $1.75

THREE PROPHETIC SCIENCE FICTION NOVELS, H. G. Wells. *When the Sleeper Wakes, A Story of the Days to Come* and *The Time Machine* (full version). 335pp. 5⅜ × 8½. (Available in U.S. only) 20605-X Pa. $8.95

APICIUS COOKERY AND DINING IN IMPERIAL ROME, edited and translated by Joseph Dommers Vehling. Oldest known cookbook in existence offers readers a clear picture of what foods Romans ate, how they prepared them, etc. 49 illustrations. 301pp. 6⅛ × 9¼. 23563-7 Pa. $8.95

SHAKESPEARE LEXICON AND QUOTATION DICTIONARY, Alexander Schmidt. Full definitions, locations, shades of meaning of every word in plays and poems. More than 50,000 exact quotations. 1,485pp. 6½ × 9¼. 22726-X, 22727-8 Pa., Two-vol. set $31.90

THE WORLD'S GREAT SPEECHES, edited by Lewis Copeland and Lawrence W. Lamm. Vast collection of 278 speeches from Greeks to 1970. Powerful and effective models; unique look at history. 842pp. 5⅜ × 8½. 20468-5 Pa. $12.95